EUROPEAN AGRICULTURE

European agricultural policy is too often a tangled web of technicalities wrapped in incomprehensible jargon. Yet it deals with the most basic human requirement – food. *European Agriculture* attempts to explain the complexities of the Common Agricultural Policy (CAP), the 1994 General Agreement of Tariffs and Trade (GATT) and the changes that are being forced on agricultural policy in Europe by environmental legislation, biotechnology and political change.

Reforms of the CAP were agreed in 1992. Explaining the reform programme in detail, the book goes on to question the effectiveness of the reforms and suggests that they will do nothing to diminish the costliness of the CAP or prevent European farm production overstepping the limits imposed by the 1994 GATT agreement. Referring to a wide geographical range of European case material, the author challenges the common assumption that in a world apparently short of food it makes sense to expand high-cost European farm production. It is clear that continued over-production will inevitably lead to new political conflicts with America and other major agricultural exporting nations, agricultural trade relationships that are crucial to the EU's continued economic and political growth.

European Agriculture is a comprehensive analysis of the economic, environmental and political implications of pursuing present agricultural policy, and presents a provocative commentary on the future development of European agriculture and its role in world food production in the next century.

Brian Gardner is an Agricultural Consultant and Director of EPA Associates SPRL, Brussels.

EUROPEAN AGRICULTURE

Policies, production and trade

Brian Gardner

London and New York

First published 1996
by Routledge
11 New Fetter Lane, London EC4P 4EE

Simultaneously published in the USA and Canada
by Routledge
29 West 35th Street, New York, NY 10001

Typeset in Garamond by
Florencetype Ltd

Printed and bound in Great Britain by
Redwood Books, Trowbridge, Wiltshire

British Library Cataloguing in Publication Data
A catalogue record for this book is available from the British Library

Library of Congress Cataloging in Publication Data
Gardner, Brian.
European agriculture: policies, production, and trade/Brian Gardner.
p. cm.
Includes bibliographical references and index.
1. Agriculture and state – European Union countries. 2. Produce trade – Government policy – European Union countries. 3. General Agreement on Tariffs and Trade (Organization). I. Title.
HD1918.G37 1996
338.1′84–dc20
95–25766

ISBN 0–415–08532–2
0–415–08533–0 (pbk)

CONTENTS

FIGURES

TABLES

Part I

EUROPEAN AGRICULTURE: DEVELOPMENT UNDER THE ‘COMMON AGRICULTURAL POLICY’

1

WHY SUPPORT FARMERS?

Why agriculture is still the most protected industry

> Agricultural policy in developed and developing nations is a tangle of contradictions. Throughout the world, governments have 'one foot on the accelerator and one foot on the brake' – simultaneously encouraging and discouraging increased farm production.
>
> (World Bank report 1990)[1]

The European Union is about to introduce a new policy to support motor car production. No matter how many vehicles the industry produces, the EU Commission will pay a subsidy equal to at least half production costs to help any manufacturer to sell cars in eastern Europe and Asia. Any surplus cars not bought by EU motorists or sold abroad will be bought up by the Commission, stored, allowed to rust and then sold off for scrap so that they can be melted down and then manufactured into more motor cars. . . . But wait, this is only a dream – or rather a nightmare. There is no such EU policy, nor is there ever likely to be. None the less, such a policy would be the industrial equivalent of the policy which the European Union actually does use to support farm prices.

The European Union spends over $45 billion a year on supporting farmers. Most of this money is spent on either buying up surplus butter, milk powder, beef and wheat, storing it and eventually selling it off cheap to Russia and the Third World, or else on paying traders to export it at cut rates. The purpose of this activity is to keep up farmers' prices. At the same time, this official market-rigging forces consumers to pay at least a third more for their food than they would in a free market. This aspect of the 'common agricultural policy' is estimated to cost the EU's 260 million consumers more than $100 billion a year.

Although some of the worst excesses of the CAP were moderated by reforms introduced in 1992, this system remains the basis of agricultural support in the mid-1990s – even after the reforms have been fully applied.

But it is not only the EU that protects farmers. Americans do it, Canadians do it, Malaysians and Arabs do it. According to a leading international economic agency, protecting farmers cost taxpayers and consumers worldwide more than $350 billion in 1993.[2]

But why if such policies are demonstrably crazy for motor cars or any other industrial product, are they accepted for agriculture? Why does the Union – and developed country governments all around the world – operate such apparently illogical policies for the agricultural industry?

As the European Union's now-notorious agricultural policy, despite quite radical reforms in 1992, continues its over-production and over-spending, detached observers may well ask: why? Why is it that politicians spend so much money and time on farmers, when in most European countries they have stopped subsidising steel mills, coal mines and manufacturing industry?

Why should a section of the population representing less than 6 per cent of the workforce and producing less than 2 per cent of the EU's economic output absorb so much money, so much political and governmental time and (in such a highly organised industrial society) so often be in the political spotlight? What other industry has its own government departments, its own Directorate General in the European Commission and its own Council of Ministers? As the World Bank points out:

> Agriculture is unique in economies. Farm sectors comprise highly competitive businessmen, producing a relatively homogeneous commodity for sale in a market with numerous price and quality conscious consumers. In other words, agriculture would appear to be the ideal industry in which to realise a textbook-perfect competitive market to the benefit of both producers and consumers. Considering the potential for pervasive competition and lack of market dominance by any one producer, one would think agriculture is the least likely sector of the economy to find extensive government intervention.

The answer to these questions is (unsurprisingly) complex and has deep historical and often emotional roots. Since the beginning of organised human society 'agriculture policy' of some sort or other has been a major preoccupation. The problem of the seven lean and the seven fat years that preoccupied the Pharaohs of ancient Egypt has continued to bother governments into the modern world. It will remain a major concern of world leaders into the twenty-first century.

There are two main justifications for supporting farm markets and they are closely linked and interlocking. The demand for food in developed countries is more or less static: unlike demand for motor cars, consumption of food does not increase if its price falls as productivity and production increases. Price does not therefore balance consumption with too often increasing production. Second, because agricultural production is determined to a great extent by climatic and biological factors, beyond human control, production fluctuates for reasons not affected by the working of the market. These variations can cause large increases or reductions in production which the farmer cannot predict; these in turn can create price slumps and booms which are ruinous to farmers and damaging to consumers. Governments therefore intervene to iron out uncontrollable fluctuations.

These two reasons would not on their own be enough to justify the vast panoply of support and expense devoted to agriculture. What makes governments worry most about farmers is that they produce the single most important basic commodity of vital importance to human survival: food. As the French *noblesse* learned to their cost in the eighteenth century and Boris Yeltsin's Russian government understands only too well today, maintenance of adequate food supplies is essential to human wellbeing and therefore to political stability.

Almost all governments therefore seek food market policies which will keep farmers producing, maintain adequate food supplies and ensure stable food prices. The only way that governments know how to achieve these objectives is to intervene in the market with the intention of maintaining steady prices and therefore also, it is hoped, adequate farm income levels.

To avoid the traditional cycle of shortages, soaring prices, over-production, bankrupt farmers and recurring shortage, governments have sought to establish support policies that iron out the peaks and troughs by giving farmers a steady price for their products. Until the international agricultural revolution which followed the Second World War, world markets were just as unstable as individual national markets, making it necessary to protect farmers from the price-depressing effects of cheap imports. In theory, this should have resulted in the matching of production to consumption and have ensured reasonable prices to consumers.

Now there is no longer so great a need for nations to protect their agriculture industries in order to secure food supplies; agricultural technology has ensured that farming has become a business like any other: a relative handful of people can produce all of a country's food supply. The spread of new technology around the world has also largely ironed out fluctuations in output.

Even as important an event as the 1988 North American drought caused little more than a blip in the upward trend of world food production. Despite this, agriculture still manages to maintain its right to special treatment. In most developed countries agricultural politics and agricultural politicians manage to maintain a disproportionately large role in the political system.

Other political considerations have too, over the years, obscured and complicated the relatively simple objective of maintaining food supplies. Aided and abetted by farmers' organisations, governments have pursued other objectives via their farm price-support policies: preservation of the countryside and the maintenance of rural communities are the most obvious.

These often justifiable objectives have also become overlaid with emotional and outdated motives like 'maintenance of the peasantry as the backbone of the nation' (nineteenth-century Prussia) and preservation of 'the rural patrimony' (twentieth-century France). Even in allegedly rational and pragmatic

Britain and the United States there is still an apparently irrational view that farmers are somehow different and should not be subject, unprotected, to the same economic forces that affect the livelihoods of shipbuilders, steel-workers or computer programmers.

Partly this is to do with the importance of agriculture in providing the basic human need – food. It is also strongly connected with the farmer's role in controlling such a large proportion of the physical environment. Not only are farmers responsible for almost all of the inhabitable land area that is not occupied by conurbations or industry, but they and their industry are also the backbone of the communities which live in rural areas.

This social importance of the farm sector – considerably out of proportion to the number of people involved – is what gives the agricultural sector its secondary political importance after the provision of food. In the European Union, with its initial very high proportion of the population still on the land – close to 20 per cent in the early days of the EEC6, the social significance of agriculture was probably as important as its strategic role. This social significance has since declined, but politicians still make obeisance to it.

Agriculture, in most developed countries, is also unique in having the most single-minded and effective political lobbying organisation.[3] This is most marked in the European Union and the United States. Like other 'single-issue' pressure groups such as industrial trade unions, the farmers' unions have only one main objective: to maintain the living standards of their members. Unlike the industrial groups, however, they can utilise a whole range of powerful strategic and emotional arguments for the maintenance of farmers' incomes. Farmers also have a social advantage over other trade unionists: in all developed countries without exception, they are an important part of the social structure of their regions – which are relatively under-populated – and therefore have an inordinate influence over local and national government politics.

In the EU, the influence of the agricultural constituency over the political system is exceptional. It is generally accepted that the inordinate involvement of the farm organisations in policy formulation is a major factor which has diverted politicians from any radical change and improvement in the operation of the CAP. The EU farmers' union federation, Comité des Organisations Professionelles Agricoles (COPA), for example, has constitutional rights of consultation in the European Union's policy formulating and decision-taking process – a right consistently denied to consumer organisations. As a federation and in its constituent national parts, it has considerable influence with the Council of Agriculture Ministers – the supreme body formulating agricultural policy in the EU – and with national ministers and ministries of agriculture, as well as with the EU Commission which administers and manages the policy.[4]

Indeed, many of the current considerable shortcomings of the European agricultural policy can be squarely blamed upon COPA, since it was closely consulted on the measures and mechanisms needed to support farmers when the agriculture policy was first established in the late 1950s and early 1960s. Subsequently, evidence from EU officials and working papers of the EU Commission and Council of Ministers shows that a great deal of the blame for the inadequate action taken by the Council to deal with surpluses and over-spending in the 1980s could be attributed to COPA's role in the decision-taking process.

In particular, these bodies accepted two COPA arguments on the future of agricultural production and trade which proved fatal to the taking of any decisions which would have averted the subsequent crisis of over-production and over-spending which evolved in the 1980s. Policy-makers in the 1970s accepted the European farm organisation's view, first, that future productivity increases in agriculture would not be anywhere as great as independent analysts were consistently warning and, second, that there would be a substantial expansion in international demand for agricultural commodities which would automatically reduce the budgetary pressure to control European food surpluses. Both contentions were proved phenomenally wrong. EU agricultural productivity in the major sectors continued to increase at over 5 per cent a year, while world agricultural production and surpluses rose to an all-time peak in 1985–86. As a result of the power and influence of the agricultural lobby in the formulation of the CAP, the EU has one of the most protected and subsidised agricultural industries in the world.

Of course, all governments everywhere interfere in agricultural markets. Some interfere more than others, however. In countries like Argentina, Australia and New Zealand support of farmers is minimal; in the United States and the European Union it is comprehensive and lavish; while in Norway, Finland and Japan it is, most experts would agree, excessive. There are of course variations in the way in which farmers are supported. In the United States – as in the United Kingdom in pre-EU times – farmers are guaranteed a minimum price which is normally above the price which consumers pay for their food. The difference between the high guarantee and the lower shop price is made up by a direct payment to the farmer, generally known as a 'deficiency payment'.

Under the deficiency payment system both farmers and consumers benefit: the farmer, through a guaranteed price above the world price level; and the consumer, from lower wholesale and retail prices. In a deficiency payment-based system it is the taxpayer who pays. In a country with an equitable tax system, the rich therefore pay more than the less well-off for supporting agriculture.

In the EU both taxpayer and consumer pay, the first through financing subsidies and the second through high retail prices for food. The EU system is thus 'regressive' since the less well-off pay proportionately more for food policy through higher shop prices for food.

In the EU the high price to the farmer is still – despite the 1992 reforms – sustained by the purchase of food commodities at an official floor price

– the so-called 'intervention price' – and by levying high taxes on cheaper imports to raise them to the same level as the high internal EU intervention price. Surplus produce is exported with the assistance of subsidies which allow the artificially high price of the EU commodity to be sold at the much lower world price level. Since 1992 the EU has begun to support farmers through a combination of market intervention and direct subsidies.

Support and protection for farmers is almost universal and for basically good reasons. These reasons have, however, over time, become overlaid with other much less good reasons. What varies and what matters is the method of support and the level at which that support is maintained. The European Union's problem is not that it supports farmers' incomes, but rather that it supports them with market price floors or subsidies – or a combination of both – which are too high in relation to the production costs of the efficient farmer and to market demand. This is the cause of surplus and a $50-billion-a-year bill for supporting the farmer.

This is why EU production has steadily increased throughout the twenty-five year life of the CAP[5] and why food surpluses and budgetary excesses have risen more than in most other developed countries.[6] The Union has not adjusted its agriculture policy to deal with the realities of modern agricultural technology and modern international political relationships. The justification for the EU's covert objective of super-sufficiency in food is challenged in a world where imported supplies are unlikely to be cut off by war or where if they were, domestic supplies would be poisoned by nuclear fall-out. More important, however, EU policy has not taken into account the effect on production of the advances of modern science.

Thus, while the individual incomes of the more efficient farmers have tended to increase – thus stimulating them to increase output even more – the incomes of farmers on average have tended to decline compared with the rest of the population. While the incomes of most people outside agriculture have increased steadily during the last two decades, the average incomes of farmers have tended to remain largely unchanged. In the EU countries where the largest sums are spent on agricultural support – Germany and France in particular – average farm incomes have in the 1980s declined the most.

What is most significant is that agricultural costs have risen most in those countries. Economists see this as indicating that the high prices paid for farm commodities have been absorbed by higher costs of production.[7] In particular, the high commodity prices have been 'capitalised' into higher land prices. This exacerbates the 'small farm problem' by forcing small landholders off the land:

> Price support policies make large-scale farming more profitable and provide incentives for expanded production on large farms. At the same time, they provide incentives for small farms to abandon farming. Rising

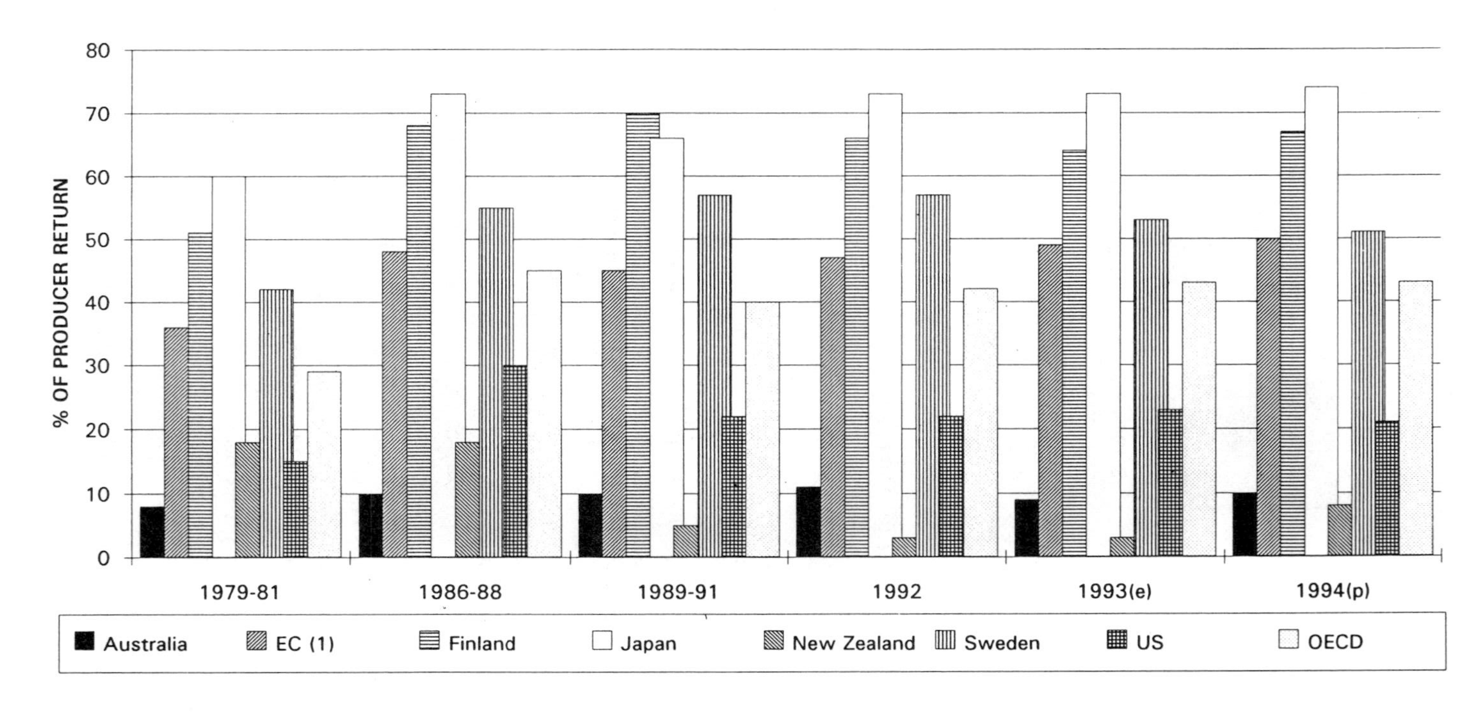

Figure 1.1 Producer subsidy equivalent OECD countries

Source: OECD

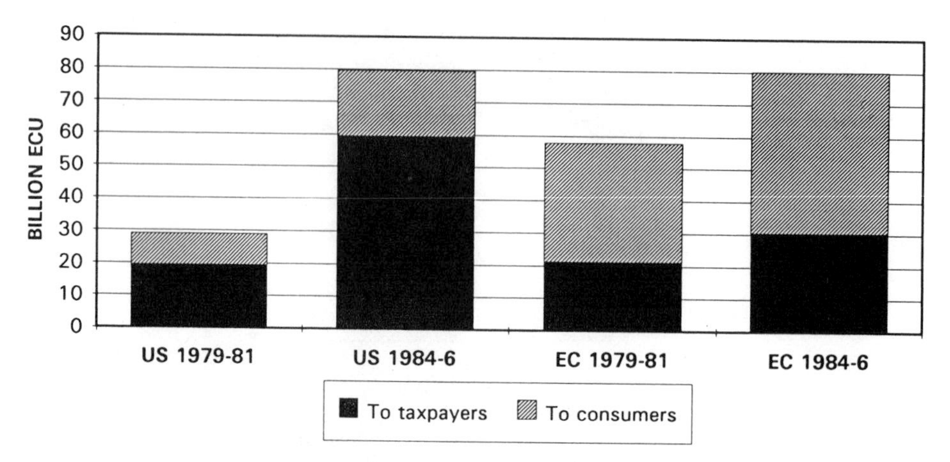

Figure 1.2 The cost of agriculture policy in the EU and the USA

> product prices lead to progressive increases in farmland prices and in farm rental values, so that it may become more attractive for small farmers to quit agriculture.[8]

In other words, the process of artificially supporting farmers becomes self-defeating: maintaining high prices does not provide adequate income support to the majority of landholders, but instead provides every incentive for them to move off the land. The inevitably larger units remain, producing yet more surplus commodities, stimulated by the excess profit resulting from support prices essentially set at a level designed to maintain the incomes of the less efficient and smaller farmers. The large farmers are harnessed to the productivity treadmill at an ever-increasing rate, while the small farmers drop out. Villages disappear and the countryside becomes more empty and desolate.

However, though all developed country governments protect their farmers, not all protect them as lavishly as the Europeans and the Japanese. In countries where agriculture is not only a mainstay of the economy, but also where it is not possible to subsidise agriculture from the wealth created in other parts of the economy, farmers have to work at close to the world price. Farmers in Australia, New Zealand and South America have to work on this basis. New Zealand is a particular example.

A country which has only been able to subsidise its farmers in the most minimal way and has always been at the bottom of the world's farm subsidy league, New Zealand in the early 1980s abolished almost all of what little support had existed.[9] The New Zealand agricultural economy is probably the most exposed to the hazards of the world market of any in the developed world. As a result, it is not surprising that successive New Zealand governments have sought to protect the agricultural sector from what was assumed to be the harmful effects of those hazards. However, in keeping with the 'new economic thinking' which became popular in the late 1970s and 1980s, the Lange Government, which came to power in Wellington in 1984, decided on a policy of general deregulation in the economy – including particularly agriculture.

In 1984 most of the government-financed props to the New Zealand farming industry were removed or their dismantling begun. The dismantling of the modest deficiency payment system – which made up farmers' losses when world market prices fell – was set in motion, interest subsidies were abolished and the Government began to remove fertiliser subsidies. In the broader economy, artificial exchange-rate manipulation, a major way of protecting farmers against world price changes, was abandoned.

The level of protection of agriculture in New Zealand is now the lowest in the developed world with the proportion of farmers' income coming from subsidies having fallen from over 20 per cent at its peak in the early 1980s to less than 5 per cent in 1990.

Unfortunately for New Zealand farmers – and the politicians who initiated the changes – this process of liberalisation coincided with other unfavourable economic factors, particularly affecting farmers. Most important were steep declines in the prices of agricultural products on world markets. Despite this, the underlying figures would suggest that liberalisation has not done serious long-term damage to the farm economy – rather, the opposite. It has to be admitted, however, that individual farmers and their families were damaged by the reforms; the number of farm families declined by 15 per cent and there were bankruptcies. It is this aspect of the impact of radical reforms which haunts European politicians.

None the less, the overall effect on the New Zealand industry was beneficial. Livestock numbers remain high, individual farm incomes of efficient farmers have not declined as far as the global figures would suggest and lower land prices – they fell nearly 60 per cent as a result of abolition of farm subsidies – will further add to the New Zealand farm industry's international competitiveness.

By the mid-1990s land values were back to the level they were at before governments began subsidising New Zealand farmers two decades ago (thus once again proving the theory first developed by David Ricardo 170 years ago that protecting farmers inevitably leads to a rise in their costs of production, which in turn leads to demands for even more protection and more subsidies).[10]

There has also been a substantial reduction in the use of fertilisers and pesticides – a very important and useful environmental bonus. What is more, the technical efficiency of the industry has also clearly improved. The New Zealand Ministry of Agriculture points out that, though the level of fertiliser and other input purchases have fallen by as much as 50 per cent, input levels are probably now at the optimal point. In other words, much of the larger fertiliser use of ten years ago was being wasted. There are also more general signs of increased productivity and efficiency.

The New Zealand example is, however, being taken to heart in Europe. Despite the traumas that the New Zealand agriculture industry has undergone, it is clear that it is possible for efficient farmers to survive without subsidies. Nor does the removal of subsidies necessarily intensify the potential threat from farming to the environment.

The Swedes, for example, took a leaf out of the New Zealand book and in July 1991 began slashing back the jungle of farm subsidies and farmer protection which had been built up in that country during the last hundred years or so. By the later 1990s, it is envisaged[11] that only the farmers who live in the far north of the country will still be subsidised – and these subsidies will be of the social, production-neutral type approved by the OECD, the GATT and other organisations dedicated to trade liberalisation and the promotion of economic efficiency.

Sweden's policy reform cut price supports, abolished quotas and shifted expenditure from market-support regimes to the improvement of the incomes and working conditions of landholders in the remote northern regions of the country.

The Swedes, as so often in European affairs, are ahead of the rest of the field in accepting and acting upon the conclusion that attempting to support farmers' incomes through subsidised market-support regimes is second only to backing horses as a way of throwing money down the drain.[12]

The reforms were a response to underlying discontent in Swedish society with the inefficiencies of agriculture policy as it developed during the 1980s. Despite increasing budgetary expenditure, farm income gains have not been significant and food prices increased faster than the prices of other consumer goods. Sweden's 1985 'Food Policy Resolution' reinforced earlier developments along familiar lines: high internal market prices maintained mainly by a high import threshold with the revenue from variable levies and excess production taxes being used to finance the export of surpluses – a system eminently designed to force the consumer to foot the bill, via high food prices, for agricultural excess.

The exportable surplus of some major commodities grew as high as 60 per cent of total production. Sweden's 'producer subsidy equivalent' (PSE) – the proportion of farmer's income coming from frontier protection and state subsidies – as measured by the OECD, has been among the highest in the world: over 70 per cent in the late 1980s.

The new policy, accepted almost unanimously by the Swedish Parliament in 1990, is designed to abolish surplus production, eliminate export subsidies and redirect support to social payments, diversification and land conservation.

An innovatory feature of the new Swedish policy is provision for direct government contracts with farmers to encourage environmental objectives. The objective of the new policy is that 'farmers should only be paid for goods and services for which there is a demand'. These goods and services will also include, apart from normal agricultural commodities, 'publicly funded services and production which are required for the fulfilment of objectives related to the environment, food security and regional policy'.

It is this interaction of agriculture with the environment which is becoming of increasing importance in the European Union. It is now at last being accepted by politicians that the combination of high EU support prices and the application of modern technology which they have encouraged has been responsible for imposing considerable strains on the rural and total environment.

Increasing applications of nitrogen to arable crops and the intensification of livestock farming in particular are now the major causes of excess nitrogen levels in drinking-water supplies; intensive agriculture has without doubt destroyed large areas of habitat for increasingly rare wild animals, birds and insects. Increasingly, under pressure from the environmentalist movement, politicians are being forced to incorporate environmental considerations into agriculture policy decisions.

The politicians are also now much less convinced that protecting farm and food markets is the way to protect farmers and to maintain their incomes. In recent years, the failure of the CAP to protect the incomes of the majority of smaller farmers or to arrest continuing rural depopulation has convinced leading EU politicians that the policy must be reformed.[13]

The problem for the Union, however, is that the CAP sets farm support prices – since 1993, the market support plus direct subsidy – at the high level theoretically needed to bolster the incomes of the less-than-efficient majority. But the problem is that no matter how high the price level, it is seldom high enough to maintain a decent income for the small and inefficient. It is simply mathematically impossible for it to do so. A French hill farmer marketing perhaps ten beef cattle a year worth $1500 each adds little to his income if the price is increased by 5 per cent; a 2000-hectare wheat farmer would, however, gain close to $12,000 from such a price increase.

At last convinced of the inadequacies of the CAP, the European Commission in 1991 made radical proposals for reform. These included the significant reduction of market manipulation and support for cereals and the use of direct payments to small farmers, to compensate them for the inevitable loss resulting from support price cuts, and possible reductions in output quotas. There were recommendations for the reduction of excess production in the dairy and beef sectors, but these were to a great extent excised by the Council of Ministers in the subsequent passage of the proposals through the legislative system.

The beginnings of a new policy based on these principles was, however, agreed by the EU Council of Ministers in 1992 and is being applied over the 1993–96 period. Though a combination of market intervention and direct subsidisation, it still maintains producer returns at too high a level and will continue to encourage surplus production of the major commodities (see Chapter 7). The subsequent GATT Uruguay Round Final Agreement of April 1994 will, however, significantly reinforce these internal EU reforms.

In its July 1991 proposals,[14] the Commission admitted the CAP's major flaw: that it is a system which bases support of farmers on the amount produced and thus inevitably stimulates production growth and encourages intensification of production.

What had to be changed, the Commission emphasised in its 1991 proposal, was the system of support dependent on market price guarantees. Because support was based on the volume of production, it therefore concentrated the greater proportion of support on the largest and most intensive farms.

The Commission's proposals were unequivocal on the objective of EU agriculture policy in the future:

> Sufficient numbers of farmers must be kept on the land. There is no other way to preserve the natural environment, traditional landscapes and a model of agriculture based on the family farm – as favoured by society generally. This requires an active rural development policy and this policy will not be created without farmers.

The future objective as far as the Brussels policy-makers are concerned is thus quite clearly to use the modified CAP as a means to maintain the structure of the countryside, as the Commission believes the majority of people want it to be. This established an important environmental and social dimension in agriculture policy formulation which has set the pattern for the future.

2

DEVELOPMENT OF THE CAP

Farmers and European politics

> The strength of farming power has long been a major mystery to commentators and a source of envy to most other pressure groups. There are few, if any, who have so consistently and pervasively exercised political influence apparently in excess of their economic or even social importance.
>
> Sir Michael Franklin, former Permanent Secretary at the UK Ministry of Agriculture and one-time Deputy Director General for Agriculture at the EC Commission[1]

ORIGINS

In the Netherlands in the closing months of the Second World War people were forced to eat tulip and daffodil bulbs; there was little else.[2] In the ruins of Germany's bomb-ravaged cities there was even less to eat. The European Economic Community's common agricultural policy (CAP) was set up only twelve years after these traumatic events had affected millions of Europeans. It is therefore hardly surprising that what preoccupied the policy planners of the 1950s was not the surpluses of today, but ensuring that Europe would never go hungry again. Most of the men who constructed the CAP were also old enough to remember the First World War when Germany and its allies were brought close to starvation by Allied blockade.

In the years immediately preceding 1957, when the then European Community's agricultural policy ground plans were laid, there had been prolonged shortages of meat, fats and sugar in most of the European countries. The exception was France, where agriculture had made a rapid recovery after 1945 and was already by 1957 looking for new markets for its expanding farm production.

It is hardly surprising that the amalgamation of these motives should result in a European system combining internal market support to ensure food security, protection against cheap imports to ensure stable food production and the facility to subsidise exports. These requirements fulfilled the food security concerns of the majority of Europeans and the expansionary ambitions of the French agricultural interest.

The system was, at least initially, cheap to run: official buying and storage of surplus was merely a matter of evening out seasonal fluctuations in production. Because the EC6 was only about 85 per cent self-sufficient in food, exports were small. Problems began when production expanded beyond domestic consumption; stored surpluses became permanent and subsidised exports had to be increased. A policy designed to ensure food security became a monstrous and expensive mechanism which increasingly created, collected and dumped surpluses on international markets.

This happened because of the fear of the many small farmers of Germany and the Netherlands that they would be ruined by low-cost imports from France's relatively more efficient farmers.

The CAP is thus the product of the uneasy post-war alliance of the French imperative of agricultural expansion with the German desire for food security.[3] By the late 1950s it was becoming clear that French agriculture was expanding to well beyond the level of output that would ever be absorbed by the domestic market or those overseas territories which were the natural first depository of French agricultural exports.

France with its high prices (in an extra-European, international context) could only compete on world markets with the aid of subventions from the state – something which the national exchequer could at that time certainly not afford. Even though French farm prices were lower than in other European countries, they were substantially higher than the prices paid to the American, Canadian or Argentinian farmers who would be the major competitors on world markets.

For Germany, the problem was different: a continuing food supply deficit and a preoccupation with industrial regeneration. For the Germans the spectre of mass starvation not once, but twice in the previous forty years, was still too much of a reality. For the Federal Republic at this time 'agriculture policy' was as much about the reconstruction of villages and the rebuilding of the rural areas as about increasing food production. Indeed, wise men in Bonn saw that in fact these two objectives were probably in conflict.

The problem was made worse by the fact that Germany had lost what had been traditionally its most productive and large-scale farming areas, now parcelled out in Moscow's east European satellites. Resolving the problem of how to concentrate the national effort on re-establishing German industrial and economic pre-eminence without either significantly increasing domestic food production (believed impossible at that time) or becoming dependent upon possibly unreliable overseas suppliers was to form the basis of not only an industrial, but also an agricultural alliance with France.

The relationship was deceptively symbiotic: the exchange of the two types of good in which both had comparative advantage to mutual benefit. It was also not how things would work out, as we shall see. The establishment of the CAP at that time was thus an integral part of the expansion of the European Coal and Steel Community into the European Economic Community.

The other four countries who became the members of the European Community of Six were willing collaborators in the creation of the CAP and for similar reasons. Italy, too, had suffered its share of privation during the war and post-war period – guaranteed access to food supplies at reasonable prices and protection from cheap imports from outside Europe was an important incentive. The Belgians and Luxemburgers had similar attitudes to food-supply security as their German neighbours, while the Netherlands took a similar view to France: better a guaranteed market with one's neighbours than the vicissitudes of the world market.

In the subsequent years the EC6 constructed a complex set of mechanisms for the support of agricultural markets and for the establishment of an import threshold which would allow imports of grain and other foods to be regulated in relation to the level of domestic production. Though unstated in the Treaty of Rome, or anywhere else, the objective was the achievement of self-sufficiency in food – the provision of as much as possible of the food supply from the production of the farmers of this new-founded 'common market', with minimal dependence upon the outside world.

National attitudes played an important part in the evolution of the mechanisms employed. Direct intervention in agricultural markets had been an important feature of German agricultural policy since the establishment of the first Reich in 1871 and the subsequent drift towards agricultural protectionism during the depression of the 1880s.[4] Germany followed France too into the establishment of substantial tariffs against food imports in the 1890s as the prairies and plains of North America began to pour their cheap and apparently endless bounty eastwards into European markets.

By the end of the First World War Germany had developed the system of state responsibility for food supplies through the establishment of state buying and storage of supplies direct from the farms or first-hand merchants. Ironically, however, Germany did not achieve the food security it desired during the First World War: heavy dependence on imports for agricultural inputs and the bleeding of agricultural resources by the army – mainly the loss of men and horses – led to rapid deterioration in food self-sufficiency and near starvation of the civilian population. Equally ironically, Britain, with its *laissez-faire* pre-war food policy, rapidly increased its food production with fewer resources during the war.[5] Britain in fact maintained its *laissez-faire* policy for agriculture until the arrival of the Lloyd George coalition government at the end of 1916; British agriculture was effectively not put on a war footing until later in 1917. Subsequently, however, there was considerable expansion in the output of cereals and the other major food crops: production in 1918 was over 50 per cent greater than at the outbreak of war [6]

German government control of food supplies was further developed under the Nazis.[7] France too developed a system of state regulation of grain supplies between the wars.

The competitive threat from France had already been clearly recognised by the farmers of the other EC6 member states in the post-Second World War period. Intensive lobbying, particularly in Germany, and many months of wrangling in the new Council of Ministers of the infant EC6 led to a compromise agreement, which, while accepting at least in principle the idea that common prices should eventually be set at the level of the efficient producer, nevertheless insisted upon a theoretical transition period to this ideal stage.

During the first period of development of the CAP, 1960–72, the planned prudent price policy which would bring about the alignment of prices on the French level was largely abandoned. None the less, the ideal of improvement in the efficiency of farms and infrastructure to allow the efficiency of farmers in the other parts of the Community to be adjusted to the French level – or at least to the level of the best French farmers – was retained.

There were of course many small and inefficient peasant farmers in France who were less well off than their German counterparts (hence the origin of the Anglo-Saxon myth that the CAP was a device for protecting 'inefficient French peasants').

It was to deal with this 'small farmer problem' that the remarkable plan of Dutch politician and EC Commissioner for Agriculture, Dr Sicco Mansholt[8] was launched. Mansholt's objective was to shift surplus labour off the farms and bring all poorer farms up to the same 'standard' of a family farm capable of providing an income equal to the regional industrial average. Although the 'Mansholt Plan' was never adopted as Community policy, the idea of the average family farm as the 'norm' upon which policy should be based was generally accepted. It is still regarded as the policy pattern by senior Brussels officials – including the architect of the 1992 CAP reforms, Ray MacSharry who was EC Agriculture Commissioner from 1989 to 1992, and Jacques Delors, Commission President throughout the crucial 1982-94 period of Community development.

The main objective of the original CAP was, however, to support farmers through the market rather than by direct subsidisation – with prices that were aimed to be high enough to maintain incomes of even the least efficient. The result was that by the early 1970s, efficient farmers in the five EC countries other than France were enjoying even better profits than in the 1960s, while those in France had the incentive of a price level higher than they had ever expected. Most important, these price levels were substantially greater than the level necessary to keep efficient farmers in business. To make matters worse, in the years after 1970, prices were increased even further.

The stage was thus set for the surplus-producing, budget-bursting farm production expansion of the 1970s. At this point, the British, finally and far too late,[9] partially overcame their post-Second World War isolationism, as well as French opposition, and joined the Community. At the same time, two important agricultural exporting nations, Ireland and Denmark, also joined.

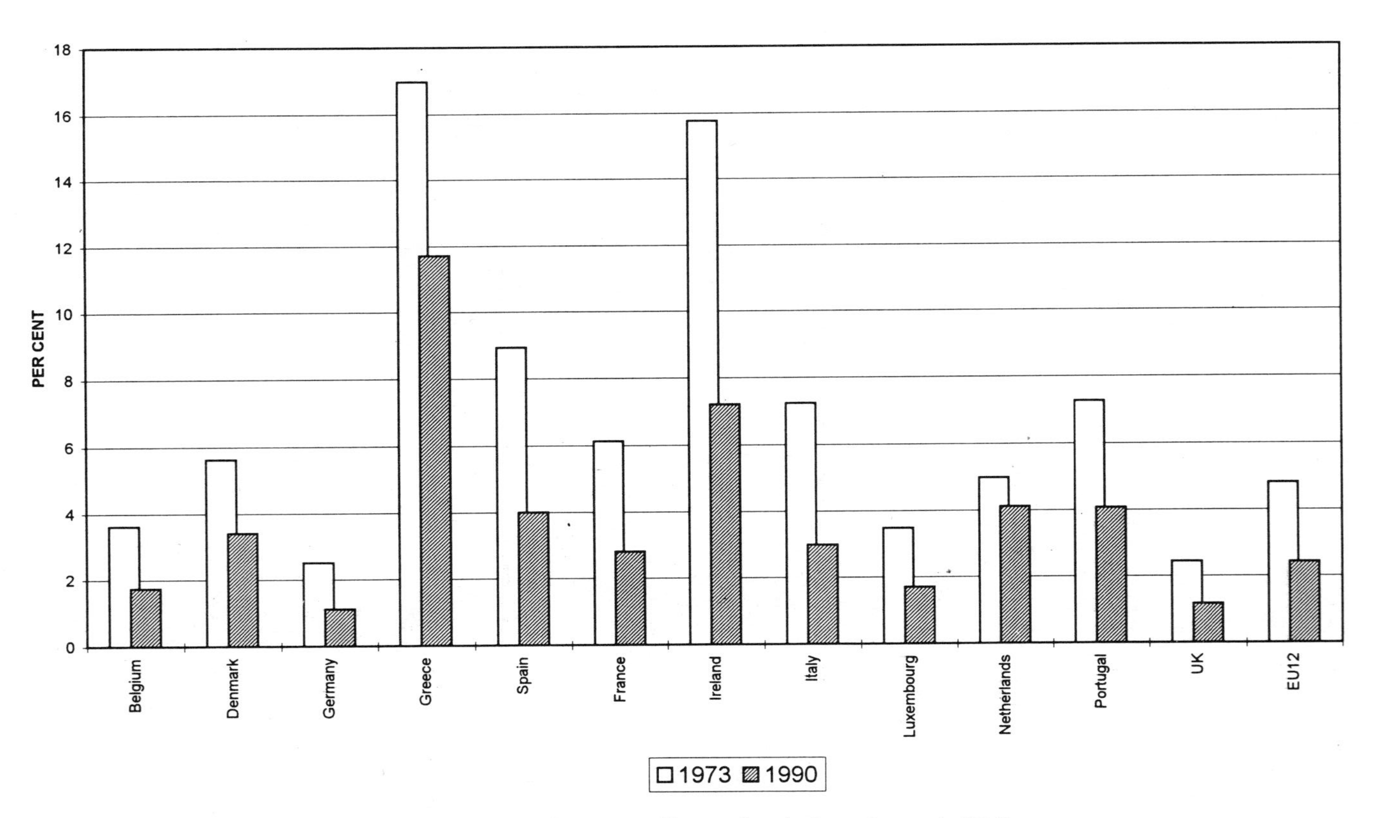

Figure 2.1 Share of agriculture in total GDP

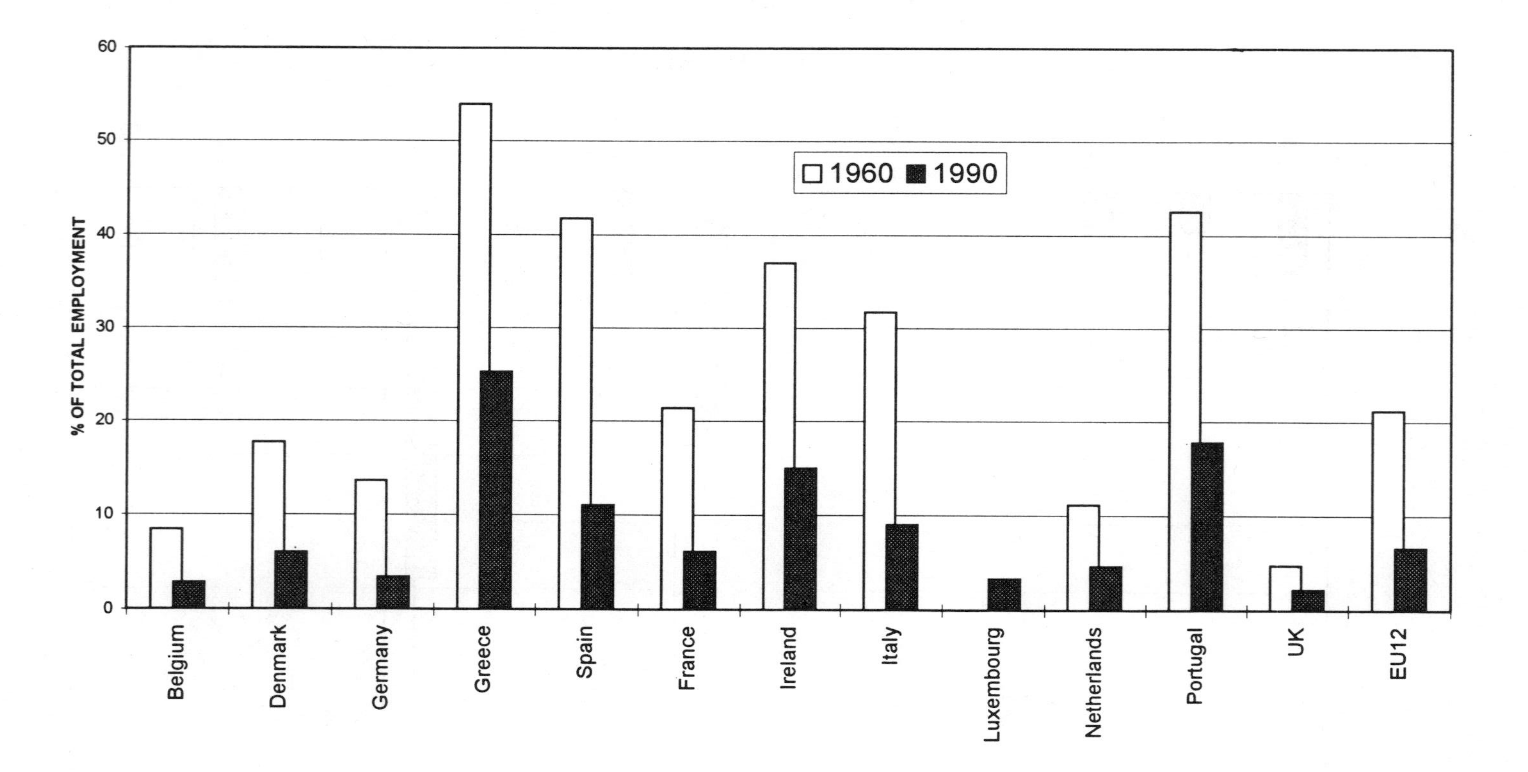

Figure 2.2 Agricultural employment

THE ROLE OF NATIONAL ATTITUDES IN THE FORMATION OF THE CAP

France

Contrary to popular (Anglo-Saxon) opinion, the general French view on agriculture policy is pragmatic.[10] Although there are wide divergences of view on the objectives of agriculture policy which vary according to political position, commercial interests and, to a lesser extent now, religion, it is generally accepted that agriculture is important in the economy and in the European Community.

Though there is still a strong attachment to the idea of the small family farm as the backbone of France's vast rural areas, this largely sentimental vision in the mind of the average Frenchman has tended to wane with the detachment of the majority of French families from land ownership and the rise of an urban generation which has finally lost its connections with its peasant roots.

In the agriculture industry itself and in its political constituency[11] it is accepted that agriculture has been modernised and that, certainly in the crop production sector – where the greatest concentration of 'agribusiness'[12] operators is to be found – France has the most competitive agriculture industry in western Europe. The 'French peasant' of popular Anglo-Saxon myth is only a feature of the remoter and less exploitable regions of the country; he and his family earn more from industries related to tourism and agriculture, as well as from regional and national subsidies, than they do from agricultural production.

With or without the CAP – and to some significant extent without it – French agriculture has been modernised during the last three decades. This process has been stimulated by migration to the towns, by the break-up of the traditional forms of land ownership, by deliberate government policy and through the dynamic effect of an expanding agricultural export industry, largely stimulated by the EC's subsidised farm export policy.

In the two decades between 1960 and 1980 no less than 690,000 landholdings in what might be described as the typical peasant grouping of between 1 and 20 hectares disappeared; the 10- to 20-hectare category declined by 290,000 in the same period. This amounted to a two-thirds reduction in all units of less than 50 ha. The number of larger than 50-ha holdings more than doubled in the 1960–80 period. Between 1960 and 1985 the number of farms in France declined by 50 per cent, peasant holdings of less than 20 ha declined from nearly 2 million to less than 800,000, and by the early 1980s fewer than 200,000 holdings were receiving more than 50 per cent of total national agricultural income.[13]

In contrast with Germany, it has, since the 1950s, been neither French government policy nor a generally desired objective of policy to preserve the peasantry. Both government and farmers' unions have since the 1950s accepted that what is generally known as the '*troisième agriculture*' – based on the unviable, too small peasant holding is condemned to disappear. Much as French politicians may still appear to pay lip-service to the needs of the small family farm – particularly in key constituencies in areas such as Brittany – the major preoccupation is with the exploitation and development of what is seen as France's vocation as an agricultural exporter. Throughout the period of the Fifth Republic there were a series of government inquiries and initiatives designed to assess the realities of French agricultural development, to speed up the process of modernisation and to exploit what became known as France's '*petrole verte*' – its vast capacity to produce and export wheat and other major farm products.

In 1969 the Vedel Report[14] realistically spelt out the likely future development of French agriculture: the maintenance of the traditional peasant structure was impossible and the number of farmers would have to be reduced to 700,000 by the mid-1980s. Throughout the 1970s a host of new legislation was brought into play to speed up the modernisation of agriculture, through farm amalgamations, the setting up of farm advisory organisations and to improve agricultural education and to stimulate efficient marketing – particularly for exports.

Much of this legislation was consolidated into the Loi d'Orientation of 1980. Although the 1980 law devoted attention to the concept of 'income parity' for farmers with their urban counterparts, the main thrust was towards modernisation and the further stimulation of the agricultural export industry. A particularly important part of the 1980 legislation was designed further to improve the laws governing land tenure and to counteract the debilitating effect of partible inheritance (the legally binding equal division of an estate among all the direct descendants of the deceased) on the agricultural industry.

Under the Socialist governments that have ruled France for much of the 1980s there has been no notable change in the general approach to agriculture. As in most – probably all – European countries, the approach to agriculture tends to be consensual and crosses party boundaries; in France this phenomenon is accentuated by the perceived benefit of agriculture to the national economy. In general the attitude can be summed up as generally mercantilist: the encouragement of the expansion of exports, the exploitation of all national and Community means to discourage imports, while at the same time actively intervening in the industry to encourage greater efficiency in both production and marketing.

Germany

In order to follow the unique development of modern West German agriculture policy and German agriculture it is necessary to bear in mind the importance of rural interests in the liberal Federal Democratic Party (FDP) and the key role of the FDP in the post-war politics of the Federal Republic of Germany.[15] The FDP has consistently held the balance of power – 10 per cent of the vote but 20 per cent of ministerial portfolios in the Federal Republic. In the 1950s and early 1960s the party was dominated by small marginal farmers and the small-town *petit-bourgeoisie*.

The professional civil service background of many German politicians is one of the major reasons for the highly consensual view on major issues – particularly agriculture. The preoccupation is with utilising rather than fighting over the acquisition of power. This 'social market' approach to German politics applies as much to agriculture as to the rest of the politico-economic spectrum.

Modern attitudes to agriculture policy in Germany date back to the formation of the concept of the 'social market economy' in the 1950s and 1960s – although underlying principles clearly originate from the nineteenth century. As the German economy recovered from the Second World War and moved into its post-war expansionary phase, the general economic debate centred on the relation of enterprises and individual workers to the market and the extent to which government intervention could be justified. To a great extent, it was decided that market forces should predominate in the industrial sector if Germany was to remain competitive. The same principle was not applied to agriculture.

Throughout the post-war period, the principle that agriculture had particular economic and social problems, indicating that it could not be subjected to the full pressure of market forces, was generally accepted by politicians of all parties – even including the essentially urban-based Social Democratic Party.

Throughout the immediate post-war period the powerful, main – and effectively only – farmers' union, the Deutsche Bauenverband (DBV) was able increasingly to argue that the farmers were bearing the brunt of what they saw as the immediate post-war 'cheap' food policy. In the two immediate post-war decades, Germany was a significant net food importer, with low tariffs allowing food to be imported at low prices relative to German farmers' costs. The greater prosperity of the urban population and the increasing power of the DBV gave rise to the view that German farmers should be protected from this foreign competition. As prosperity increased, expenditure on food came to occupy a much smaller proportion of household incomes. With more than 90 per cent of farmers adhering to the DBV, the union was a force that had to be reckoned with by all parties. The DBV view that farmers should be excluded from the effects of the social market gained currency.

The DBV increasingly pressed the line upon politicians – with mounting success – that agriculture policies should be aimed at giving farmers parity in income with industry. Most important, not only was this income parity to be maintained, but at the same time, rural population levels were also to be sustained. Given the obvious conflict between these two objectives, it was clear that it could only be reconciled by protection and subsidisation. The implication of the transition period (1967–71) for the cereals policy of the CAP – that German prices would have to fall to the French level, galvanised the DBV into action to counteract this likely trend.

The switch in voting by farmers from the CDU (Christian Democratic Union) to the FDP in the period immediately after the EC cereals policy agreement in 1964, convinced politicians of the seriousness of the agricultural unrest and of the importance of the farm vote for the future development of German politics – despite the apparent small number of rural voters. The CDU/CSU (Christian Social Union) coalition of the 1960s could not have retained its share of the vote in the 1965 and 1969 elections without the agricultural vote.

None the less, the 1969 election saw the end of the coalition and the rise of the importance of the FDP as the lynch-pin party. This development in Germany was crucial to the later development of the CAP – as had been Germany's role in the formulation of the CAP prototype. At the same time, it saw the emergence of the Agriculture Minister Josef Ertl, who not only articulated the desires of the DBV into policy, but also came to typify the essentially paradoxical attitudes of Germany to the CAP.

Ertl could also be justly blamed for laying the foundations of the problems of surplus, monetary distortion and complexity which were to dog the CAP in the 1970s and early 1980s and which are still too present and obvious today. Ertl's canniest move was continuously to promulgate the view that the lack of 'monetary cohesion' (essentially, the tendency for the Deutchmark to revalue continuously against the French franc and other weaker EC currencies) was the main obstacle to a true 'common agricultural market'. Until such cohesion could be achieved, he was never to cease to argue, German farmers had to be protected from competition from other EC farmers whose greater competitiveness arose not, according to Ertl, from their greater efficiency, but essentially from currency advantages. In the case of France and Britain, efficiency *was* more important than monetary differences.

Ertl's first notable move was indeed to gain EEC acceptance – with willing French connivance – for the establishment of 'monetary compensatory amounts' (for an explanation of this iniquitous development, see Chapter 3) to give protection against lower French prices following the 9 per-cent devaluation of the franc in August 1969. This first breakdown of the infant CAP under monetary stress, through imposition of intra-Community monetary balancing import levies, gave German farmers an effective 8.5 per-cent protection against 'cheaper' imports from France and other weak currency EC countries. The balancing of trade via monetary fluctuation, as described by the classical economists, had no place in this 'common market'.

The idea of the rural economic and social problem requiring special 'dirigiste' attention was given effective political support through a comprehensive plan for establishing a domestic German rural social policy which was in direct conflict with the then EC objective of increasing agricultural efficiency through the encouragement of larger holdings and the movement of labour out of agriculture. What became known as the 'Ertl Plan' effectively aimed at the opposite: the provision of subsidies to ensure the survival of the smallholding capable of supporting at least one labour unit – in other words, the small family farm.

During the 1970s German agriculture policy came to have two main features: the development of the rural structural policy, which was the main feature of the Ertl Plan, and the heavy dependence upon the EC intervention system as the means of maintaining market prices for farmers. Germany thus increasingly exploited the EC's agriculture policy – through the Brussels-financed intervention system – to maintain the high market prices that were necessary to protect and sustain the small-scale, high-cost farm structure which government policy had already established and was by the mid-1970s consolidating. The dependence on intervention demanded two other essentials: the continuous upward adjustment of intervention prices themselves via the EC price-setting system and the maintenance of balancing monetary compensatory amount levies and subsidies (MCAs) to ensure that the high price levels thus achieved were not diluted by cheaper imports from other EC countries.

The United Kingdom

Traditionally, Britain had maintained, first, a free-trading food supply policy and then post-imperial manipulation of its food markets, designed to supply food at the lowest possible price to its predominantly urban industrial population. Between 1848 and the outbreak of World War II this policy had developed into the exploitation of the agriculture of both colonial territories and other large-scale, low-cost producers in the new worlds of the Americas and the Pacific. With the growth in the food production of the Great Plains of North America and the development of cheap, steam-powered shipping in the latter decades of the nineteenth century, this led to the progressive ruin of large-scale arable agriculture in the British Isles.[16] After a partial recovery during the First World War, the post-war depression pushed British arable agriculture back into what had become its traditional poverty-stricken state. The subsequent sustained depression of the 1924–38[17] period did, however, persuade some politicians that some assistance to agriculture could be justified on both food security and social grounds.[18] The modest agricultural support legislation of the early 1930s – which marked Britain's first real break with the philosophy of 1846 – was the response. The Wheat Act of 1932, the Import Duties Act of 1932 and the Agricultural Marketing Acts of 1931 and 1933 guaranteed minimum prices to cereal producers via deficiency payments and allowed farmers to establish marketing schemes for major commodities with statutory powers of first-hand purchase – provided that at least two-thirds of the producers concerned favoured such a move.[19] Although this degree of protection was regarded at the time as exceptional, it was in fact minimal by post-Second World War standards: by 1938 government expenditure on agricultural support was a mere £21 million compared with the £299 million of 1971[20] and the 1995 level of £1.2+ billion (including both national government expenditure and the United Kingdom's share of EU expenditure).

There were, however, considerable misgivings about the abandonment of *laissez-faire* agriculture policies in the 1930s. Some saw a suitable compromise in using the tariff system to discriminate against non-British Empire exports and to foster increased deliveries of grains, dairy products and meat from Canada and Australasia. None the less, such protection of the domestic industry as there was remained minimal and food imports, despite the depression, actually increased by 8 per cent in the 1932-39 period.[21]

From 1938 onwards it was generally accepted that war was likely and that much greater production would be needed from domestic agriculture. In stark contrast to the attitudes of 1914, Britain's governments rapidly instituted new policies in the 1938–41 period designed to maximise production and reduce import dependency: guaranteed prices backed by deficiency payments a vigorous ploughing-up policy and the encouragement of modernisation and expansion. The arable area increased by 50 per cent and overall production increased to meet close to 60 per cent of total temperate food needs.[22]

This new support structure and the dirigiste attitude towards agriculture persisted into the post-war period; the economic straits in which Britain found itself after the war encouraged the establishment of a production-maximising, import-saving policy in 1947. In the post-war period there was also some recognition that it was necessary to stimulate agriculture in order to maintain a stable rural social structure. The result was the Agriculture Act of 1947, which effectively established support for farmers and management of the food market so that domestic production was maintained at an 'economically reasonable level' while a significant proportion of supplies of livestock products, cereals and sugar came from Commonwealth producers and other overseas low-cost producers.

Ironically, this new protection of British agriculture was not introduced by a government of land-owning Tories, but by a reforming Labour government. It has been argued that this development was part of the post-war Socialist attempt to weld all of society – including agriculture and the rural sector – into one corporatist whole. A more pragmatic conclusion is that the war had demonstrated the dangers of too great dependence on imports and that a nation bankrupted by war could save vital foreign exchange by producing more of its food at home.

The wording of the 1947 Agriculture Act was sufficiently flexible to allow the proportion of supply to be obtained from the domestic industry to be adjusted to fit political contingencies. The Act stated that the intention was to maintain

> A stable and efficient agricultural industry . . . producing such part of the nation's food and other agricultural produce in the UK . . . at

minimum prices consistent with proper remuneration and living conditions for farmers and workers in agriculture and an adequate return on capital invested in the industry.

Unlike contemporary mainland European agriculture policies, the prior objective of this post-war British policy was the provision of food supplies at 'reasonable prices'; the protection of farm incomes, at least at that point, took second place to this primary aim. This was an important difference of approach to policy design which still underlies the broad intellectual differences on the purpose of food and agriculture policy between the British and the other Europeans.

The British tended to bring with them to Brussels in 1973 the conviction that the deficiency payment was the most efficient method of supporting agricultural markets. It was not generally accepted in Whitehall that market intervention (the official buying-up of surpluses) and the variable levy (a variable tax on imports) might not be more appropriate to what was undoubtedly rapidly becoming a large and self-sufficient European food market.

Inevitably, the British attitude to the CAP and to agriculture policy in general has been considerably modified during the United Kingdom's seventeen-year membership of the Community – almost to the point of schizophrenia. On the one hand, there is the view that the CAP is essentially unreformable and that therefore the appropriate way for Britain to minimise the damage to the national economy of high food prices is to make sure that those high prices are paid to British farmers via maximised domestic production; and, on the other, there is the declared desire to reform the CAP in order to reduce EC budgetary expenditure and to reduce international conflicts arising from the dumping of EC surpluses on international markets.

It is the former attitude which has predominated. The reason for this is simple: agriculture policy is formulated by agriculture ministers. Even British farm ministers are the creatures of the farm and food industry pressure groups who maintain constant pressure on agriculture ministry officials. Once having protected itself against EC budget excesses, via the new budget burden sharing arrangements fought for and won by Margaret Thatcher in 1980, the United Kingdom has done little more than pay lip-service to the objective of fundamental CAP reform and has been a substantial contributor to the increase in production and surpluses in the 1970s and 1980s. While presenting an image of active reform for general public assimilation, British agriculture ministers – just like their German counterparts – have exploited the 'green currency' system (see Chapter 3) to insulate largely their own farmers against the main effects of the ECU-denominated price freeze that has operated through most of the 1980s.

The United Kingdom none the less, likes to regard itself as instrumental in the recent development of the CAP. Margaret Thatcher's criticism of the profligate spending on the EC agriculture policy in the 1979–83 period and

	1985	*1991*	*1992*	*1993*	*1994*	*1995*
Olive oil	692	1874	1754	2302	1999	875
Fruit and vegetables	1231	1106	1262	1743	1722	1901
Wine	921	1048	1087	1666	1567	1044
Tobacco	863	1329	1272	1401	1235	1160
Total Med. products	3707	5357	5375	7112	6523	4980
EAGGF total	18372	32386	32107	35352	36465	36897
Non-Med. products	14665	27029	26732	28240	29942	31917
Mediterranean %	20.18	16.54	16.74	20.12	17.89	13.50

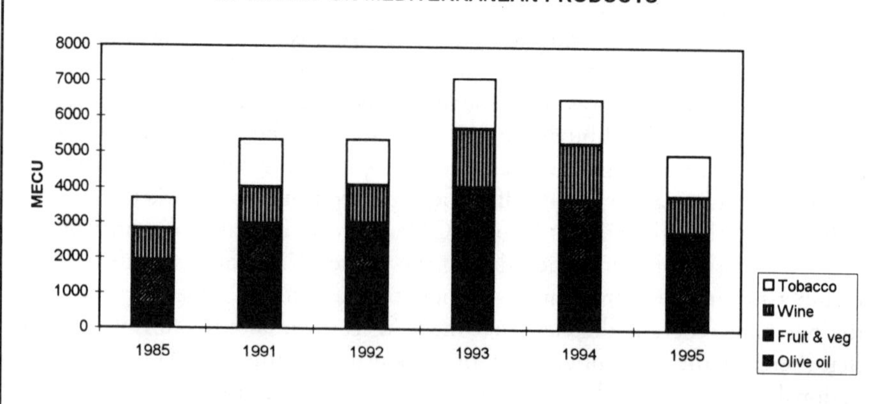

MEDITERRANEAN PRODUCTS' DECLINING SHARE OF BUDGET

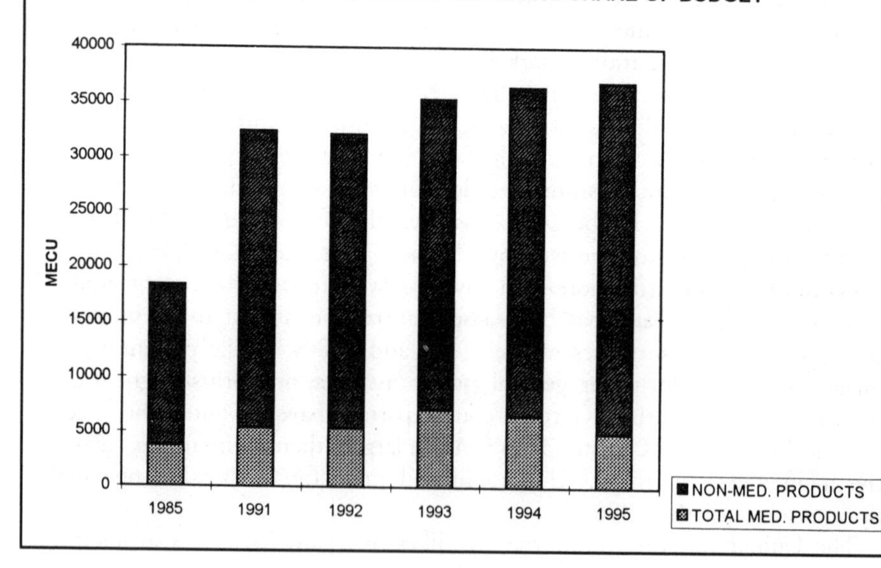

Figure 2.3 EAGGF budget support of 'Mediterranean' products

her demands for 'compensation' for Britain's perceived lack of benefit from the CAP did of course have the effect of concentrating European minds on the need for reform.

The CAP was, until the reforms of 1992, still none the less largely the policy designed by the EC6 in the 1960s. A few 1980s 'add-ons', such as quotas, budgetary 'stabilisers' and limits on intervention stockholding, gave an outward appearance of reform, but by the early 1990s the beast was still the basic money-squandering, surplus-creating and trade-distorting monster which emerged from the Franco-German alliance in the 1960s. It took the need to forge freer industrial trade relations with Japan, the newly industrialising nations of the Pacific Basin, the United States and the rest of the developed world – epitomised by the 1993–94 GATT agreement – to force the non-agricultural section of the European political spectrum to make the agricultural constituency begin to mend its profligate ways. The 1991 MacSharry proposals for reform were the result.

It remains to be seen whether the 1992 reforms (Chapter 6) which resulted will really reduce waste and achieve the desired objective of a more efficient food, agriculture and rural environment policy.

The Mediterranean countries

Conflict between the northern and the southern members of the European Community is a continuing undertone of EC agriculture policy formulation. The EC Commission's attempts in the 1980s to cut support for citrus fruit and to impose further limitations on support for the wine market, as well as the changes in the dried-fruit market support system, are seen by the southern countries of the Community as specific attacks on the incomes and living standards of Mediterranean producers. Farm groups and governments in these countries complain that support for the mainstream northern products – particularly dairy products – remain largely untouched.

The south/north dispute is however a symptom of a much deeper and longer-running resentment which underlies relations between the Mediterranean areas and the north of the Community. The essence of the arguments is the relatively small proportion of the EC farm budget (EAGGF – European Agricultural Guarantee and Guidance Fund) expenditure which has habitually gone to the Mediterranean areas. This is combined with the contention among southern producers – particularly in Italy – that they have had to bear the brunt of the cost of not only the enlargement of the Community in the 1980s, but also of the increase in food import concessions given to non-EC Mediterranean countries.

3

HOW THE EUROPEAN UNION SUPPORTS AGRICULTURE

One May morning in 1989 investigators from the EU Court of Auditors were amazed by the apparently crazy activities of dockworkers in the bustling port of Hamburg. With conveyors at both ends of the hold of the freighter *Kapitan Danilkin*, they were loading wheat out of the ship at one end – and back in again at the other. The wheat, from France, was being unloaded for a brief touchdown on German soil before being reloaded for export to Russia. But there was nothing crazy about the reward for this activity: an extra export subsidy bonus from the Brussels farm fund of around $17 a tonne – in addition to the normal subsidy for exporting surplus EU wheat to countries outside the Union which was then $98 a tonne.

The reason for the bonus for this apparently pointless exercise? An exceptional surplus of wheat in the German market had prompted Brussels to offer an extra export bonus to clear it into the international market. Luckily for crafty exporters, however, the Commission had failed to specify that to qualify for the extra $17, the wheat had to be *German* wheat – it only had to originate from a German port before being exported to a non-EU country. Wheat was therefore being brought from all over the EU to Hamburg to be unloaded and smartly reloaded to gain the 'German origin' necessary to claim the extra subsidy.[1]

There are two important points about this event: first, it was not in the strict legal sense fraudulent; second, it illustrates that most of the agricultural price-supporting activities of the EU Commission are concerned with paying subsidies to the agricultural trade rather than to farmers. Farmers' incomes in the Union are generally supported not by direct payments from EU coffers, but by paying traders to store, process and export surpluses and potential surpluses. This has to some extent changed since the reforms of 1992.

Until 1993 the European Union seldom supported farmers by paying them direct subsidies from the public purse. Until that year, the 30+ billion ECU which the EU was spending each year on farm support went to two main mechanisms which maintain minimum market prices for the major food commodities – grains, milk, beef, sugar, oilseeds, wine, and fruit and vegetables. The first is direct buying-up of surplus commodities by the EU authorities at minimum official prices to maintain a minimum 'floor' in the market. This is supplemented by the payment of subsidies to traders to sell surpluses of the main commodities on the generally lower-priced international markets – export subsidies or 'restitutions'. These 'EU subsidies to farmers' of popular myth are thus nothing of the sort, but rather subsidies to food buyers, processors and traders who are paid by the EU to store, process and export the produce from European farms and thus, indirectly, maintain prices for producers. For example, EU cereals traders were paid an average of over 75 ECU a tonne to sell EU wheat on the international market in the 1978-92 period.[2] Since 1993, however, an increasing proportion of expenditure also goes on direct subsidies to compensate farmers for the reduction in these market price supporting mechanisms; expenditure on intervention and export subsidisation still remains substantial, though.

In May 1992 the EU's Council of Ministers agreed reforms of the CAP which put more emphasis on direct payments to farmers (see Chapter 6). Support of farmers in the EU is, however, still likely to be heavily dependent upon export subsidies until at least the end of the 1990s.

The system works like this. Each year the Council of Agriculture Ministers, with the assistance of the EU Commission, sets minimum prices for all the main foods. These minimum prices are in practice the 'intervention' prices at which the EU is committed to buy up produce whenever it is offered by food traders. If prices are falling and likely to fall lower than this intervention price, then it is obvious that traders will sell to intervention rather than onto the open market. In good times, when prices are high, sales to the official intervention stores are small or non-existent. But the tendency, since the mid-1970s, of farmers to produce more cereals, more beef, more butter, more milk powder and more wine than EU consumers can eat and drink has meant that for most of the 1978–93 period intervention purchases were extensive and stored surpluses consequently large. The market price support system works as follows. Farmers sell their wheat, beef cattle, milk or sugar beet to traders and processors who eventually sell the finished flour, beef steak, butter or refined sugar to wholesalers or retailers who are the unavoidable conduit to the consumer. Under the EU market support system, there is generally no direct payment of subsidy from the EU authorities to the farmer for the commodities which he produces. The authorities do, however, support the secondary markets for beef carcasses, wheat, butter, milk powder and sugar through what is known as intervention buying, thus indirectly supporting the price which wholesalers and dealers pay farmers.

For all of these products the authorities are normally legally committed to buying any consignments offered to them by traders at the annually fixed 'intervention price'. This price is therefore the floor in the market for all of these basic food products, since no trader will sell on the open market at less than the intervention price because he knows that he can always sell at the minimum intervention price to the EU authorities.

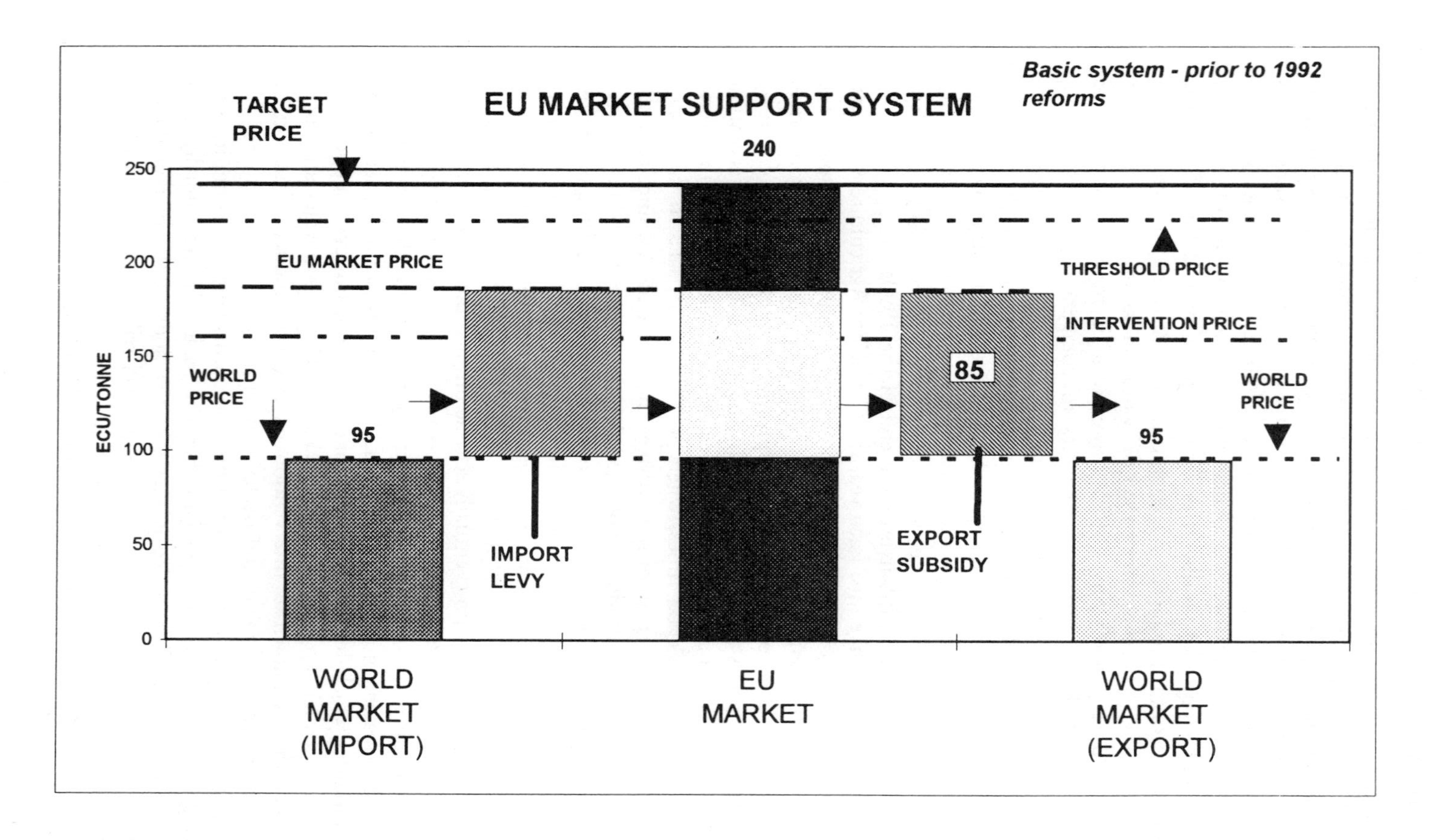

Figure 3.1 The EU agricultural market support system

In theory the 'intervention system' is supposed only to provide temporary relief to food commodity markets: taking surplus off the market in time of seasonal surplus and feeding it back into the market when the temporary over-supply has disappeared. For most of the time, also in theory, market prices should be above the intervention level and there should be no buying-in by the EU authorities. There should also be no permanent accumulation of surplus in the EU stockpiles either, since surpluses accumulating at seasonal highs should be dissipated when production falls. Indeed, this was how the system worked in the very early days of the CAP.

The problems began, however, when EU farmers began to respond to high – too high – EU prices and to produce more than the domestic market could absorb. Depression of market prices became, instead of a temporary and infrequent feature of EU food markets, permanent. Once production had reached and exceeded the level of full domestic consumption – the point of 'self-sufficiency' – the EU intervention price for heavily surplus products like butter, milk powder, sugar, beef and wheat became *the market price* and the growing stocks in the official intervention stores became permanent.

They also became an increasing burden on the EU budget and the Union taxpayer. At its worst in 1986, buying up and storage of butter alone cost the Union taxpayer more than $2 billion; in that year the Council of Ministers had to 'borrow' $4.5 billion outside the normal farm budget in order to dump some 1.3 million tonnes of butter on the world and domestic markets. About a tenth of this record stockpile was so badly deteriorated from having been in store so long that it was inedible – even by animals – and had to be disposed of by being melted down and burnt in power station furnaces.

But, it may be reasonably asked, how does this system, despite its expense, manage to succeed in keeping up the price to the consumer? What is to stop food wholesalers and supermarkets from buying cheaper supplies which are undoubtedly available from abroad?

The answer is simple: the Union effectively bans the import of any food at less than the minimum import price set by the Council of Ministers. This so-called 'threshold price' is literally that – the minimum price level at which any foreign produce is allowed to flow into the Union market – and this minimum import 'sill' is higher than the EU internal market price. Cheap wheat from Australia, cheese from Canada or beef from Argentina has to pay a 'variable levy' that always equals at least the difference between the very much lower – usually at least a third lower – world price and the high Union import threshold. With butter and beef, the import sill or threshold is so high that it is impossible for any produce to be sold into the EU market from abroad at a price which could be competitive with domestic production. In the case of some products such as butter the import threshold system is effectively a complete ban on imports. Small amounts of these basic commodities are allowed into the Union under special import concessions

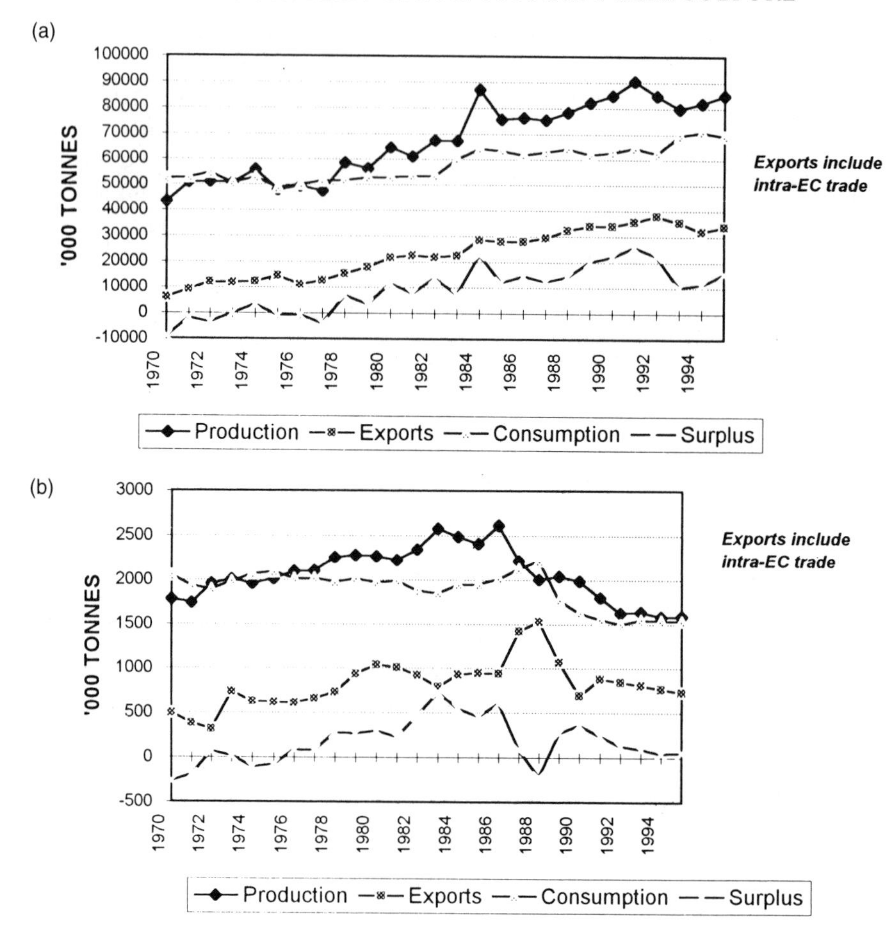

Figure 3.2 EU growth of surplus and exports: (a) wheat; (b) butter; (c) sugar; (d) beef

for developing countries or old-established trade deals like the butter import arrangement between the United Kingdom and New Zealand. In no case, however, does the concession amount to more than 5 per cent of EU consumption.

Just to complete the explanation: the intention of the combination of domestic market intervention and the high import threshold is to push the price to the EU farmer towards what is known as the 'target price' – a price above the intervention price but below the import threshold. In practice, however, the price target is seldom achieved. Because production of all the main commodities so far exceeds the amount that the domestic market can absorb, the market is wholly dependent upon the intervention system to maintain the minimum price to the farmer.

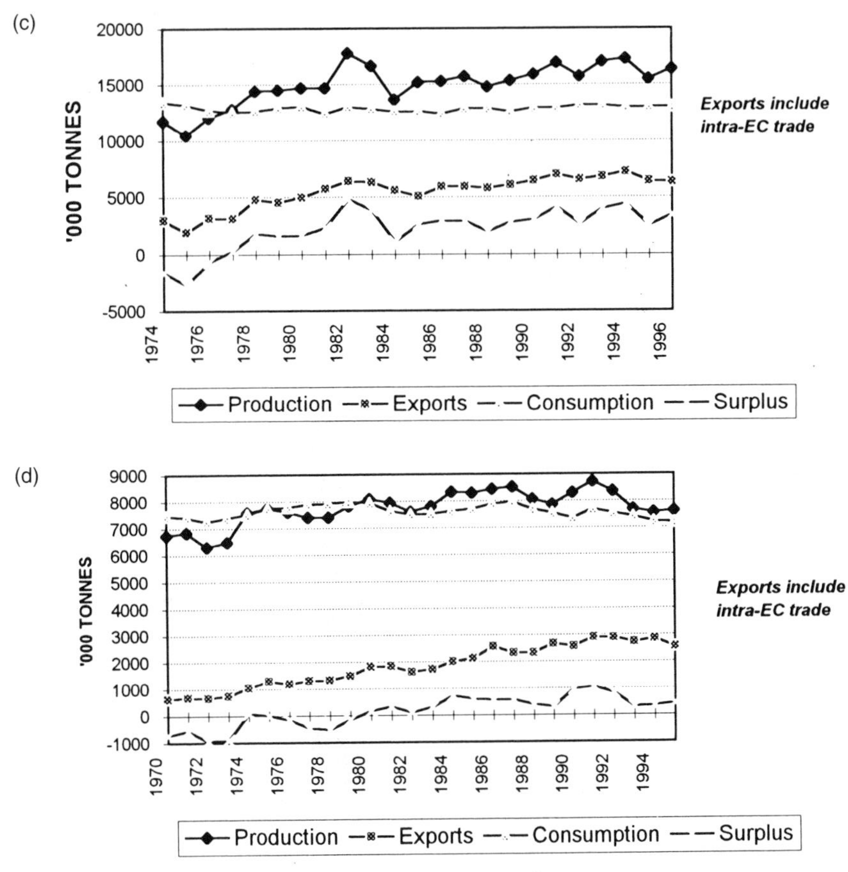

Figure 3.2 (continued)

But, again in practice, for several of the major commodities the farmer does not even actually receive the intervention price which should be the minimum which he ought to receive. Why should this be so? There is a relatively simple and practical reason. The EU's intervention system is designed to support the farmers' price indirectly, by supporting wholesale rather than primary markets. The EU intervention stores do not buy up milk from the cow, cattle 'on the hoof', sugar-beets from the field or small lots of wheat from the combine harvester; they buy only the butter and milk powder processed from the milk, carcass meat from slaughtered cattle, sugar refined from the raw beets and large 100-tonne lots of cleaned and graded wheat. This means therefore that the farmer can only sell to intervention through the intermediary of the dairy, the butcher, the sugar refiner and the grain merchant. The EU system thus guarantees the price of the middleman – not the farmer.

Clearly, the price which the farmer receives for his produce must be influenced by the price which the trader receives from the EU-bolstered market. The

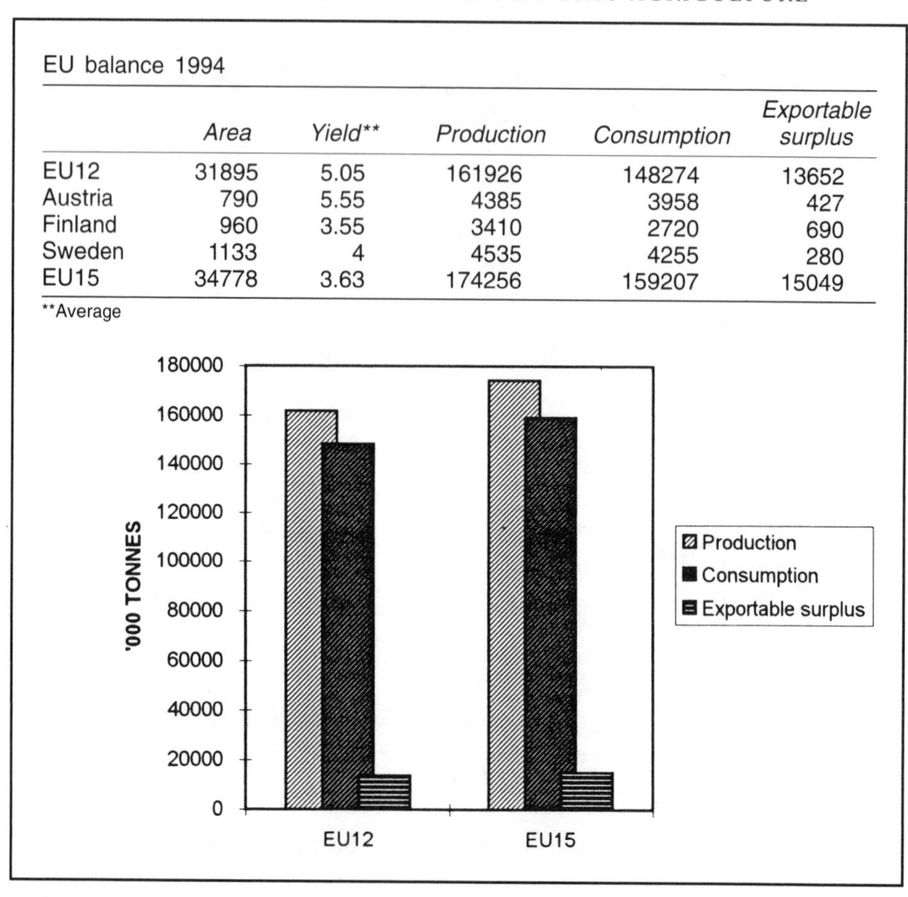

EU balance 1994

	Area	*Yield***	*Production*	*Consumption*	*Exportable surplus*
EU12	31895	5.05	161926	148274	13652
Austria	790	5.55	4385	3958	427
Finland	960	3.55	3410	2720	690
Sweden	1133	4	4535	4255	280
EU15	34778	3.63	174256	159207	15049

**Average

Figure 3.3 EU15 grain balance

important truth about the EU market support system is that it is just that: a means of supporting traders' prices on the main food markets rather than of directly supporting farmers' incomes. Indeed, close analysis of payments out of the EU agricultural market support budget (EAGGF Guarantee) reveals that until the changes made in 1992, more than 70 per cent of disbursements went to traders in food rather than to the primary producer.[3]

At its worst, this system can be seen to support the trade while in practice failing to support the farmer.

Take beef as an example. Farmers produce beef cattle which, when they have reached their optimum size and condition, are sold off the farm for slaughter. These animals go to markets where they are sold by auction, or direct to dealers who then sell them to butchers who slaughter them and process the carcasses into the neat and tidy steaks, joints and offal which appear on the retail slab. Sometimes farmers sell directly to butchers. There

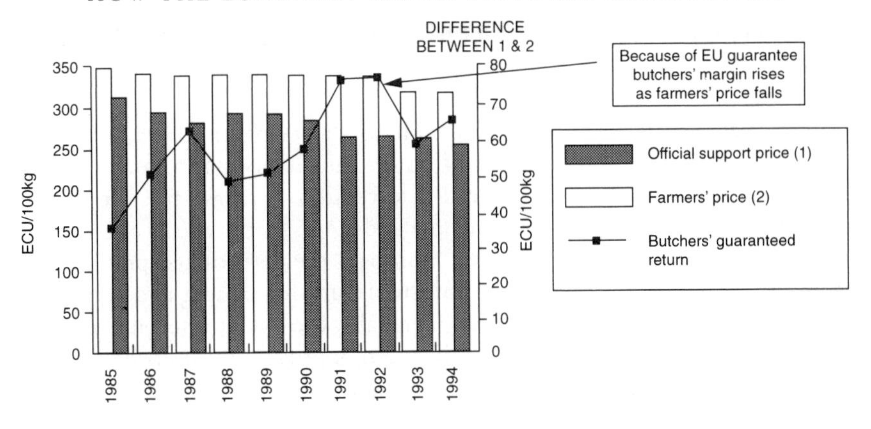

Figure 3.4 The EU beef market: how intervention supports the butchers' profits

Note (1) Official buying-in (intervention) price
(2) Average market price of beef animals ex-farm EU12

is, however, one important point to note about this process: the beef producers are many and the butcher buyers of cattle relatively few.

In a market free of official management this is not generally a good thing from the producers' point of view: in times of over-production, which occur regularly – on average every three and a half years – the many producers compete against one another to sell to the few buyers, thus depressing the prices which they, the producers, receive. Bearing in mind that the butcher and meat processor have a guaranteed price set by the EU system, what happens in times of over-production in the EU-supported market? Beef farmers' prices fall as they compete against one another to sell to the processors, but the processors' profits do not fall. This is because, as the price of their raw material – the beef cattle – falls, the gap, the profit, between this lower input price and the fixed, guaranteed EU price for the processors' output, the dressed carcass, increases.

Remember that the butchers' price cannot fall because it is guaranteed by the EU intervention system, into which surplus carcasses and jointed meat can be sold if a high price cannot be obtained on the open market. The absurdity of this system is illustrated by the undeniable fact that the profits of butchers and meat processors increases as the price of beef 'on the hoof' falls. This would not happen on an open market without interference from the EU authorities, because the meat handlers would have to reduce the price at which they sell to the retailers in order to clear the market. In the EU system, on the other hand, both the farmer and the consumer lose out: the farmers get lower prices than they should, consumers pay higher prices than they need, while at the same time forking out an annual 5 billion ECU for the storage, degradation (deep-freezing beef halves its value) and international dumping of some 500,000 tonnes of beef a year. Even after the 1992 reforms, substantial quantities of beef are being and will still have to be exported with substantial subvention from the taxpayer.

There has also been an increasing bill to the taxpayer for this market-support operation. To buy a tonne of butter into intervention, for example, not only costs its 3200 ECU official intervention purchase price, but also approximately 250 ECU a year to handle and store. At the end of the storage period – on average, at least eighteen months – the deteriorated product has to be either sold off for catering or industrial purposes or sold at a very large discount on the international market. Generally, the sale of a surplus product out of intervention takes place at a price no more than a third of the original buying-in price. The cost of taking a tonne of butter off the open market therefore typically costs the EU authorities, or rather the taxpayer, approximately 3450 ECU. As the level of EU production of the major foods in excess of consumption has increased, so have the amounts of surplus that have had to be bought up at intervention.[4]

The bill for internal market intervention operations (that is, not including the cost of export subsidies) has climbed steadily through the 1970s and 1980s: from a mere $3.5 billion in 1972, to over $17 billion in 1978 to over $22.4 billion by 1986. By 1994 it had reached more than $45 billion.[5,6]

But this direct market intervention is only half of the market support system. In addition to excluding imports of cheaper food from the international market by the imposition of heavy import taxes (levies) which vary so as always to equal the difference between the high EU support price and the lower world market price, the Union also pays subsidies to traders to sell high-priced EU products on the usually lower-priced world market. Like the import levy, the export subsidy varies with fluctuations in world prices. As the world price falls against the fixed EU internal price, so both the import tax and the export subsidy increase.

This means that cheap imports can never undercut the internal market and that EU traders can seldom be undercut on the world market by traders with cheaper produce from non-EU countries. No matter how cheaply the third country operator may be selling, the EU always follows a falling international price with bigger export subsidies. Not surprisingly, therefore, since the world's food markets have been in a state of over-supply for most of the lifetime of the CAP, the bill for export subsidies too has risen steadily as EU surplus production has increased.

When the EU produced less than its own needs of the main foods in the early 1970s, export subsidies – paid only on cereals – cost the Brussels farm fund less than $700 million a year. By 1990 more than $19.6 billion was being spent in dumping not only surplus cereals but also beef, butter, milk powder, sugar and wine on international markets.

The modifications of the CAP introduced during the 1980s put some limitation on the intervention buying activities of the EU farm support

Table 3.1 The degree of self-sufficiency among member states, 1976 and 1990

	Total cereals		*Degree of self-sufficiency in total cereals (%)*		*Self-sufficiency levels in 1990*			
	Production 1990	*Utilisation 1990*	*1976*	*1990*	*least*	*%*	*most*	*%*
	million tonnes							
BLEU	2.2	4.2	39.0	54.3	maize	5	barley	82
DK	9.2	6.1	104.3	143.8	maize	0	rye	182
D	26.5	24.7*	80.1	107.1	maize	58	rye	132
GR	5.4	4.5	88.9	104.4	barley	77	rye	131
E	18.8	20.1**	n/a	100.0	maize	75	barley	117
F	57.5	25.4	153.1	221.2	oats	111	wheat	252
IRL	2.1	2.1	67.4	102.5	rye/maize	0	barley	137
I	16.6	20.4	71.2	78.9	barley	67	maize	91
NL	1.3	4.4	24.9	30.1	maize	0	wheat	52
P	1.3	2.8	n/a	49.8	barley	36	oats	103
UK	22.6	19.5	65.4	113.6	maize	0	barley	139

* The figure for utilisation in Germany has been estimated since figures for 1991 and onwards are for the unified Germany
** The apparent discrepancy between the data for Spain can only be explained by different data sources.

Source: European Commission

authorities. There were no equivalent limits on the subsidising of exports. There was however, little reduction in the growth of surpluses. Therefore, instead of pilling up in intervention mountains, surpluses were, in the later 1980s, increasingly exported onto the international market. Despite the claims of politicians, while surpluses of cereals and oilseeds continued to increase, their presence was obscured from the public gaze by being exported: icebergs floating around the world's trade channels, rather than mountains growing in the Union's own back yard.

Even those areas of the food market where production has been checked by policy changes still produce extensive excesses. Despite the introduction of dairy quotas in 1984, the Union still produces about 15 million tonnes of milk that cannot be sold in the EU market – that is the equivalent of about half a million tonnes of butter plus a million tonnes of milk powder. It is for this reason that, despite ten years of milk quotas, the Brussels farm fund still has to spend over 4.5 billion ECU a year on the disposal of dairy product surpluses.

From 1993 cereals and oilseeds growers will get part of their return from direct subsidy payments. In the main, however, EU payments are still made to processors, storers and exporters of food. Brussels' hand-outs to the food trade are not only an important stimulant for the food industry, but also form a very important element in the profits of big food combines.

The bigger the company, the bigger the gain from the CAP. The large amounts of money involved and the complexity of the regulations also makes it a very fertile grazing ground for fraudsters and manipulators. The EU's Court of Auditors has, in recent reports, indicated that at least as much as 10 per cent of the farm budget – around 3.5 billion ECU a year – is lost to fraud, the bulk of it undetected.[7]

Leaving aside actual fraud, big operators can also manipulate the system perfectly legally to suit their own financial and political ends. Probably the most outstanding example of this were the activities of the late Jean Baptiste Doumeng, the self-styled 'Red Millionaire' and his French Inter-Agra company.[8] Over a period of some fifteen years, between 1970 and 1985, Doumeng exported millions of tonnes of surplus wheat, beef, milk powder and butter as well as millions of litres of wine to the Soviet Union. To do so he persuaded the EU Commission to part with hundreds of millions of ECU in export subsidies; so lavish were these subsidies that at one point the Union was effectively paying Doumeng to give butter and beef to the Soviets; EU taxpayers were not only paying an export subsidy equal to the original price at which the Commission had bought in the surpluses from the EU market, but were also paying the transport cost to the USSR. A large part of the profits of Inter-Agra went to boost the funds of the French Communist Party.

Second, because the CAP operates largely through the market rather than through direct subsidies, it benefits large farmers rather than the smaller landholders. The reason is obvious: the income of a farmer producing 5000

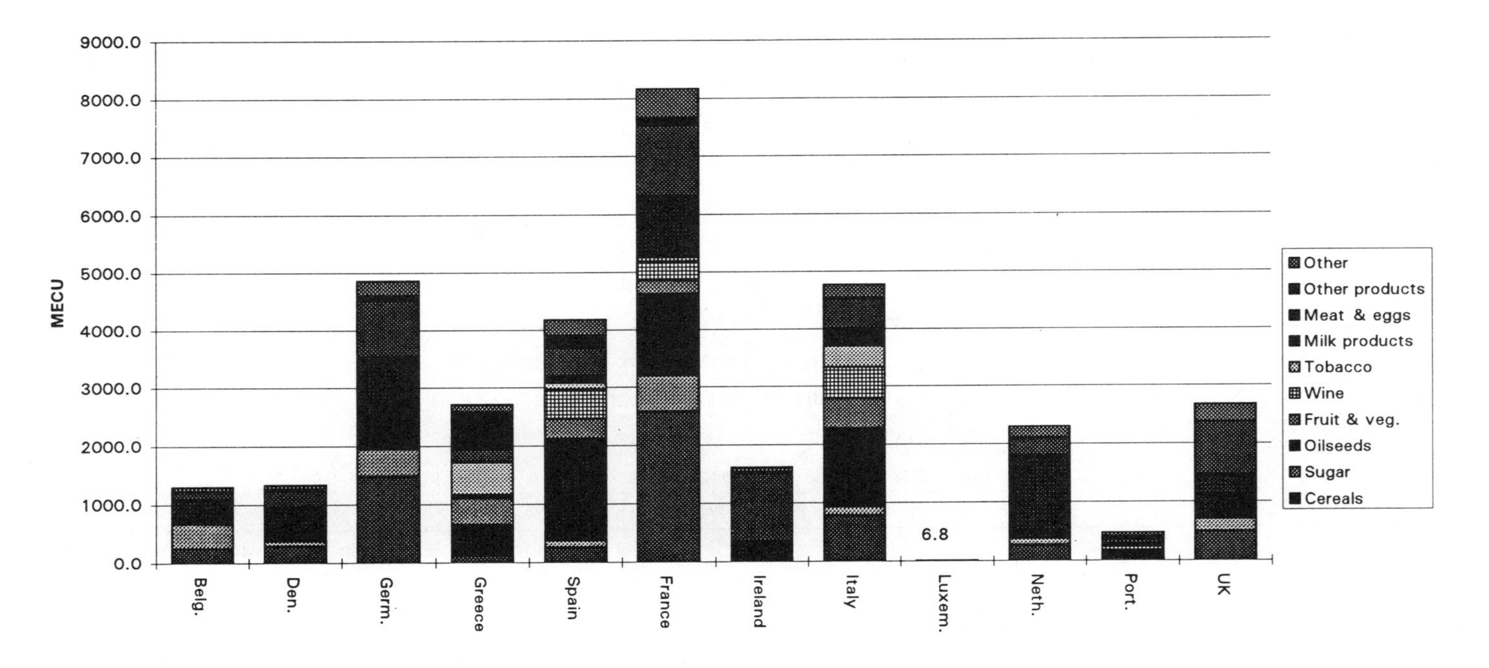

Figure 3.5 EU agricultural support cost by country and commodity

Record of payments made in 1993, by member state and by market (**MECU**)

	Belg.	*Den.*	*Germ.*	*Greece*	*Spain*	*France*	*Ireland*	*Italy*	*Luxem.*	*Neth.*	*Port.*	*UK*	***Total EU***
2.1 Cereals	254.6	296.8	1474.3	135.9	257.5	2583.9	39.6	777.2	0.1	267.8	45.2	497.0	6629.9
2.2 Sugar	416.5	84.4	470.7	10.8	116.3	608.3	9.0	149.9		110.8	7.9	204.1	2188.7
2.3 Oilseeds	59.9	146.6	771.2	505.2	1743.9	1426.8	10.8	1349.9	1.1	113.3	89.5	398.0	6616.2
2.4 Fruit and vegetables	20.8	3.7	30.2	447.1	347.0	237.6	0.3	508.9		32.3	36.0	8.2	1672.1
2.5 Wine	0.0		15.2	59.9	503.6	326.1		545.1		1.6	57.4	0.7	1509.6
2.6 Tobacco	10.0		43.9	548.0	109.6	68.9		370.2			14.4		1165.0
2.7 Milk products	324.1	430.4	746.3	11.6	110.8	1053.1	290.7	296.9	2.6	1248.6	43.7	337.4	4896.2
2.8 Meat and eggs	127.2	272.8	968.6	210.5	498.6	1227.2	1153.7	514.3	2.4	311.3	76.3	915.8	6278.7
2.9 Other products	11.6	23.6	73.8	654.0	221.2	138.9	13.4	19.3	0.1	15.7	28.2	11.0	1210.8
2.10 Other costs	74.2	74.1	267.9	135.6	279.7	496.8	88.8	241.6	0.5	198.0	80.3	306.9	2244.4
Total	1298.9	1332.4	4862.1	2718.6	4188.2	8167.6	1606.3	4773.3	6.8	2299.4	478.9	2679.1	34411.6

Figure 3.5 (continued)

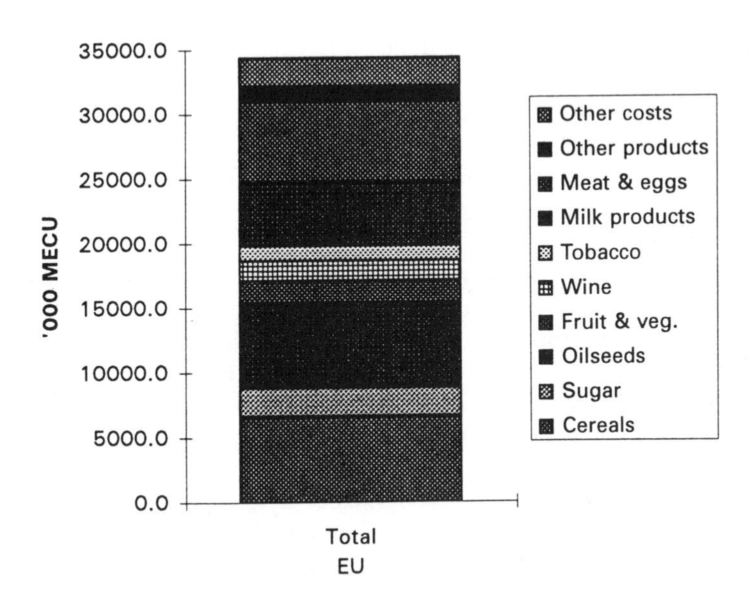

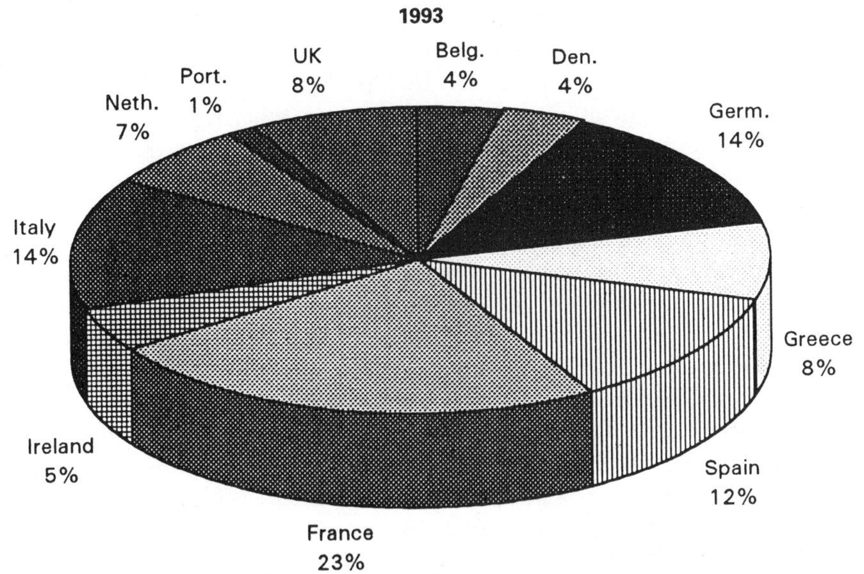

Figure 3.5 (continued)

tonnes of wheat a year worth around 160 ECU/tonne receives a very substantial boost when the Council of Ministers raises the support price by 5 per cent. The income of a small sheep farmer in the Auvergne, producing fifty lambs a year worth around 3000 ECU, hardly changes if the price were to increase by 5 per cent. This is one of the reasons why the CAP has signally failed in one of its major objectives of supporting farmers' incomes: the support of small farmers.

In fact, the gap between the incomes of the large farmers becoming richer and the small farmers becoming poorer has widened significantly over the lifetime of the CAP.[9]

Third, the CAP creates serious inequity in the sharing of the budgetary cost between the EU member states. The EU budget – and therefore the CAP – is financed through the income from levies on imports of agricultural and other goods from third countries and from direct contributions from each country based on its annual value-added tax yield. At first sight, this would appear to be a straightforward and equitable system: the larger the economy the greater the contribution to the central fund. The problem is, however, that because of the dominance of agricultural spending in the EU budget,[10] agricultural exporting EU countries pay less and food importing countries pay more. This is because the pay-back from the Brussels farm fund to an agricultural exporting country for export subsidies is larger than its pay-out for levies on imports of food from third countries. This meant that, prior to the adjustments to the budgetary system following the 1980 squabble sparked by UK Prime Minister Margaret Thatcher's demand for at least £1 billion compensation for this distortion, every one of the EU countries with the exception of Germany and the United Kingdom drew more from the Brussels budget than they paid in.[11]

Payments for agricultural export subsidies and other CAP activities were in each case greater than the payments into the budget under the VAT-based contribution system. Although Germany was prepared to tolerate this distortion, the United Kingdom was not.

The EU subsidised loading and unloading of the *Kapitan Danilkin* described at the beginning of this chapter – with the same cargo of wheat being unloaded at one end of the ship and reloaded at the other via an officially approved German silo in order to claim a lucrative supplementary EU export subsidy – has now entered the extensive anthology of CAP lunacy. But it is not an isolated event. More recent reports from the Court indicate something more insidious and important: that the steady manipulation of the EU's labyrinthine regulations in order to milk the maximum in subsidies from the Union's farm budget is not only common, standard practice among the EU's food processors and agricultural traders, but is also accepted by such companies as standard practice.

In 1992 the Court of Auditors published the result of a special audit on the subsidy-claiming activities of two of the largest dairy companies in the Union.[12] This demonstrated that these companies were habitually claiming

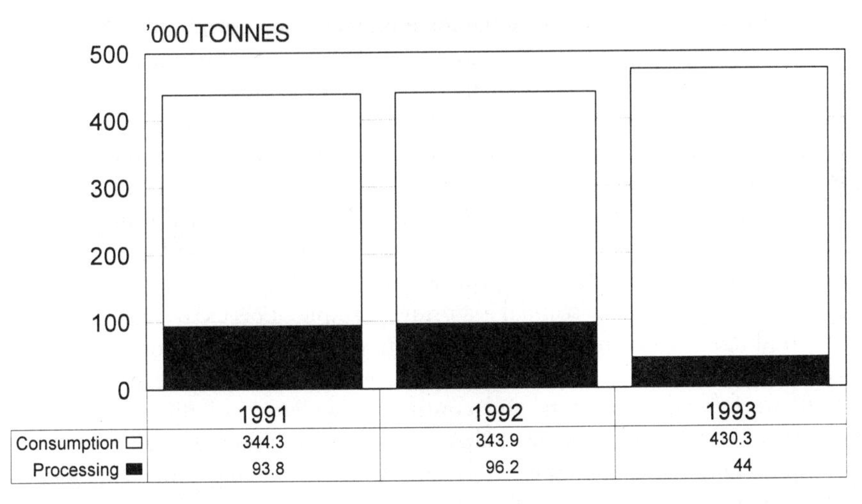

Figure 3.6 EU butter: subsidised consumption (tonnes subsidised annually)

export subsidies on high-priced cheese – entitled to a large export subsidy – while actually delivering low-price cheese which should have been entitled to only a small export subsidy.

These examples of exploitation of the excesses of the CAP, quoted by the Court of Auditors, serve to illustrate at least two important points: first, the more that the authorities meddle with markets, the more convoluted the means necessary to sustain that manipulation; and, second, companies and individuals do not need to be criminal – merely ingenious – to make a great deal of money out of the Union's agriculture policy. This would suggest that there is something fundamentally wrong with the policy itself.

To take the last point first: the policy is so full of absurd and lavishly financed mechanisms designed to support every small or large, real or imagined weakness in the market that it is often possible to make large profits merely by moving goods around rather than by actually buying and selling them. The important political point about what the Commission aptly describes as this 'exploitative trading' is that it is not illegal. It may be immoral in terms of the waste of resources involved and the loss of taxpayers' money, but such activity is completely within the law; indeed, the irony is that it needs to be completely legal, with all the correct stages of the process officially recorded, to be effective. The 'borrowed' lorry loads of butter that progress round the Union collecting monetary subsidy differentials when the currency movements are in the right directions, the lorry loads of wheat – the same lorry loads – that journey back and forth across the same frontier without being unloaded in order to collect a favourable monetary subsidy margin, must be duly observed and recorded by customs officials if the 'trader' concerned is to collect his multiple payments from the EU authorities.

The loss from the EAGGF Guarantee budget from real fraud – the export subsidies claimed on goods that do not exist, subsidies claimed on goods of a higher quality than those actually being exported or processed, the subsidies paid out for non-existent olive trees, for the grubbing of phantom orchards, for the retiring of imaginary cows – all these scams may well, as is expertly estimated, account for up to 10 per cent of the 36 billion ECU a year laid out on agricultural support, but the amount lost through legal – but immoral – 'exploitative trading' must at least equal and probably far exceeds the loss through fraud proper.

The *Kapitan Danilkin* scandal – a prime example of perfectly legal 'exploitative trading' – arose from the fruitless attempts of the EU authorities not only to maintain the myth of a common agricultural market, but to sustain it in the face of continuing and continuous widening of the gap between the most valuable and the less valuable currencies within the Union.

This is a problem which the EU Commission has had to grapple with almost since the beginning of the establishment of a common agricultural policy: how to maintain a common price level when currencies were constantly fluctuating. The problem was first set when the French franc was devalued by 8.5 per cent in August 1969. This meant that, when combined with a smaller subsequent revaluation of the West German Deustchmark, exports of grain from France to Germany cost 11.5 per cent less than they did before the devaluation and revaluation of the two currencies. German farmers complained that their market price was therefore being undermined by cheaper grain pouring in from France. The Bonn Government demanded action at the EU level.

To solve this problem the EU Commission invented 'green money' – no, not a creation of Edward Lear or Lewis Carroll, but a mechanism designed to sustain one single price when the values of the currencies in which that price was set were constantly changing. Just to sustain the 'Alice through the Looking Glass' impression, this invention actually *put back* the import taxes and export subsidies between EU member states which the CAP was principally designed to remove.

Henceforth EU farm prices were to be denominated not in francs, Deustchmarks or florins, but in 'units of account'. While all official ('institutional') prices are stated in units of account (post-1979 – European Currency Units, ECU), they clearly have to be converted into national currencies – pounds, francs, lira, Deutschmarks etc – before traders or farmers can be paid intervention, export processing or other subsidies. The 'agricultural unit of account' was to be converted into these national currencies at fixed 'reference rates' – the so-called 'green currencies'.

But the problem was, and still is, that farmers and traders deal in real money values denoted by the foreign currency markets – not the fictional 'green money' exchange rates set by Brussels.

So any variation from these fixed exchange rates had to be compensated for by either a levy (an import tax) or a subsidy, depending on whether the currencies involved in the transaction were depreciating or revaluing on the foreign exchange markets. In the initial French devaluation which sparked this invention, these 'monetary compensatory amounts' (MCAs) – import taxes out of France, import subsidies into France – filled the gap between the falling French franc price of wheat and the rising DM price. French wheat sold into Germany carried an export tax (levy) on its way out of France and a further import tax on its way into Germany; German dairy exports to France carried an export subsidy on their way out of Germany and a further subsidy on their way into France.

In this way, both the higher price level in Germany and the lower price level in France were maintained by the imposition of these MCAs at the frontiers. Expensive German food remained competitive on the French market and cheap French food could not undermine the German market, while at the same time prices remained the same in the Brussels unit of account. But this is just a simple explanation of the very beginning of this 'agri-monetary' system which has since become the major obstacle to a common agricultural market and to the rational exploitation of comparative advantage within the European Community (subsequently, European Union).

As farm and food trade expanded between the countries of the EC6 and, later the EC9 and ten, so the system became more complex. Since every type of food moving in trade is subject to some subsidy or other, those subsidies themselves had to be adjusted for monetary compensation. A biscuit filled with jam and cream, exported to the United States, for example, would not only carry export subsidies on the cereals, cream, milk powder and sugar which it contained, but also monetary adjustments of all those subsidies and in addition, MCA export subsidies too. Such a relatively simple product would require at least sixteen different calculations to be applied before the compound export subsidy which the exporter could draw would be ascertained.

But the green money system was soon discovered by the EU's agriculture ministers to have a much more profound and sinister use. It could be manipulated by the agricultural politicians to achieve different price levels in the member states from those indicated by the common price level. Ministers of countries with persistently weak currencies – for most of the 1970s and 1980s, these included the United Kingdom, France, Italy and Ireland – the adjustment of the green currency (the agricultural reference or exchange rate) downwards towards its depreciating average rate had the effect of raising farm price levels denoted in national currency. This meant that even though the Council might have agreed to freeze or even reduce prices – as it did with milk prices in the later 1970s and early 1980s – prices could still be increased in Britain, France and Italy by 'devaluation' of the 'green' pound, the green franc and the green lira. Farm ministers of the weak currency countries continually sought, and were granted, permission to align their green money on their continuously declining real currency values.

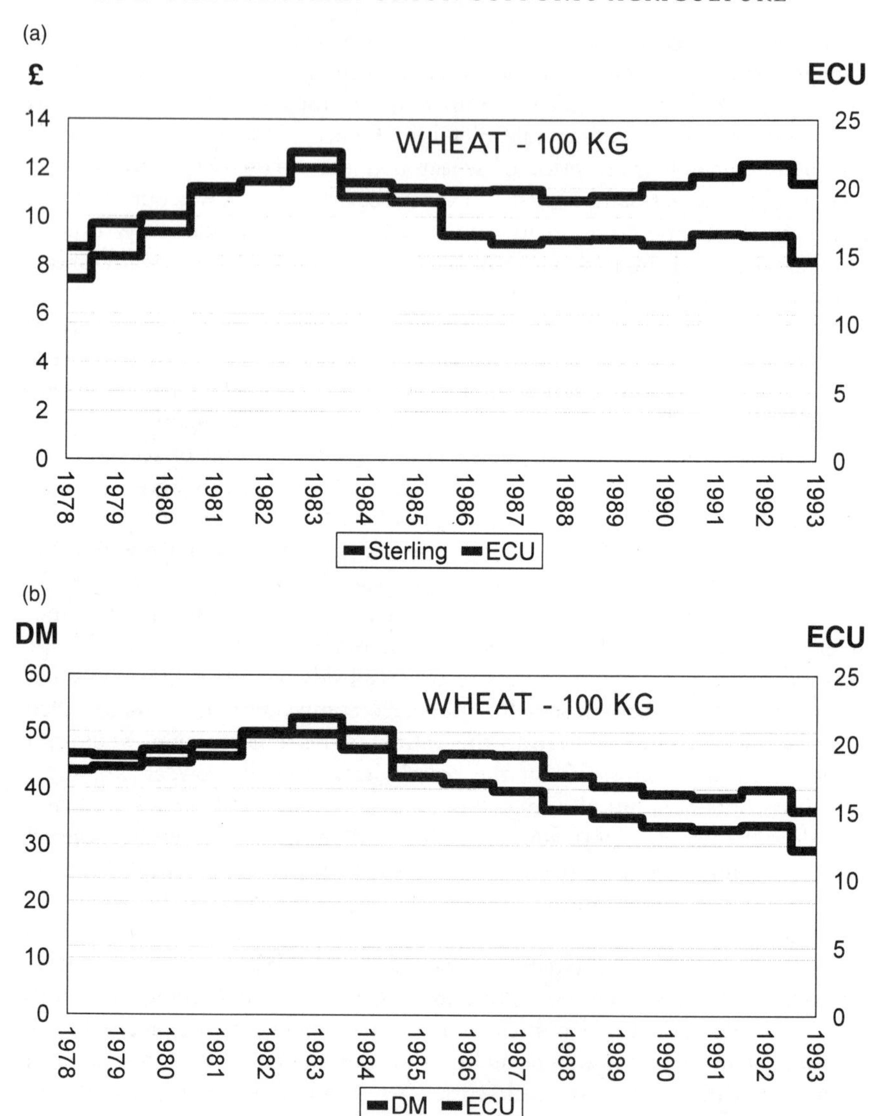

Figure 3.7 Effect of EU agri-monetary system: (a) depreciating currency (sterling); (b) appreciating currency (DM)

This process became a tremendously valuable tool with which agriculture ministers could protect their farmer constituents from the intended austerity of the EU Commission's farm pricing policies.

Thus, while most of the citizens of these countries were suffering the normal harmful effects of a devaluing currency – inflation and loss of purchasing power – farmers were protected from this trend through continuously rising prices. To make matters worse, British, French and Italian consumers had to pay for this process through higher food prices created by the green currency devaluations.

Thus while the EC practised a nominal 'freeze' on farm prices between 1978 to 1993, farm prices in the United Kingdom and other weak currency countries continually rose in national money terms. Over this period of so-called farm price freeze, average prices in national currency actually rose by close to 21 per cent.

There was also a further profoundly damaging feature of the green money system: its tendency to inflate the underlying EU food and farm price level in relation to farm commodity prices in the rest of the world. This originated from two main factors: the dominance of the revaluing DM in the calculation of the 'agricultural unit of account' – it carried a weighting of more than a third – and the revaluation of the agricultural unit of account itself to take account of the appreciation of the German currency.

Because of this, by 1978 when the European Monetary System was established, the agricultural unit of account was 13.7 per cent more valuable against the ordinary European unit of account. Conversion to the single European Currency Unit (ECU) in 1978–79 should have involved devaluation of the agricultural unit to parity with the ECU – and a consequent cut in EU farm prices of a similar level. This was firmly resisted by the Agriculture Council; instead, they agreed to the creation of an 'agricultural ECU' worth nearly 14 per cent more than the ECU proper. Once again, a privileged price level for agriculture was built into a major general change in EU monetary policy.

Throughout the twenty or more years since its creation this 'agri-monetary' system has needed a vast army of officials in Brussels and in the member states to administer it. The monetary amounts – subsidies and levies on hundreds of products had – until the system was modified in 1993 and again in 1995 – to be changed every week in order to keep up to date. At the height of the monetary disparity in the Union in the early 1970s one German official complained that he had to deal with daily telexes from Brussels which were getting on for a quarter of a kilometre long.[13]

Despite this, it is practically impossible for the adjustments of the green money system to keep pace with the changes in currency values from day to day and from week to week. Thus there is nearly always an advantage for traders to sell cereals and other commodities from weak currency countries into the strongest currency country, Germany. This is why in the heyday of the operation of intervention as the main means of supporting markets in the late 1970s and the 1980s the largest stocks of wheat, of butter, of milk powder and often of beef were usually to be found in the Federal Republic.[14]

In the later years of the CAP's development, the fact that export subsidisation has become the main means of supporting farmer's prices does not mean that the flow of commodities towards the FRG has declined – it has remained as distinct a feature of the agricultural market as it ever was. To prevent a serious dilution of the German market by the inflow from other states the Commission has, however, been forced to find ways of artificially clearing the market in that country. One such means was the special additional subsidy paid at German ports to encourage the exporting of the 'German' wheat surplus which was at the root of the *Kapitan Danilkin* incident.

The reports of the Court of Auditors[15] also make an important broad political point: although CAP surpluses persist on a large scale, they are now being more carefully obscured from unthinking politicians and journalists than in the early days of the CAP, by being exported on an increasing scale. The Union, even after the reforms of 1992, still produces substantial surpluses of wheat, milk, beef, sugar, wine and vegetable oils; it no longer holds them in official stores on the pre-1986 scale. Instead, it has since the mid-1980s been dumping them on international markets at ever-increasing cost to the EU taxpayer and the third country exporter.

As the Court of Auditors says:

> Over the past two years [the period 1988-90] the subsidised export of agricultural produce to countries outside the Union has replaced intervention storage as the principal instrument by which the Union maintains the equilibrium of the internal market. . . . Export refunds have superseded public intervention storage as the key-stone of the price guarantee system under the common agricultural policy.

This trend is what has brought the Union into conflict with other food-exporting countries. So extensive and often unconsidered has the operation of this system been that 'the value of the export subsidy ("refund" or "restitution") payable on a single export consignment will often exceed the price payable by the third country customer'. The auditors give several examples: a consignment of beef to Libya worth 18,600 ECU carrying a total refund of 34,700 ECU, 450 tonnes of butter to the USSR worth 150,620 ECU with a refund of 867,850 ECU and 2000 tonnes of milk powder to Algeria worth only 1,536,430 ECU but subsidised by the EC taxpayer to the tune of 2,515,000 ECU.

The Court picks out one, large and outstanding, example of the way in which the Commission is often manipulated, willingly or otherwise, by large companies operating in export markets: the export of butter to the Far East and the USSR. In March 1986 export licences for the export of 610,000 tonnes of butter were granted on the basis of tender. The Court's investigations clearly reveal, however, that the so-called tender was rigged by one large operator in every country of the Union where butter was available. The trader concerned was subsequently able to export the 610,000 tonnes of butter with a subsidy equal to more than 90 per cent of the value of the butter itself.

Although different companies appeared to be involved in the tenders, they all managed to offer the same (low) price in all countries. When the details emerged, 610,000 tonnes of the total 715,300 tonnes were seen to have been sold to only one buyer.

The Auditors conclude that not only has the export refund (subsidy) system 'developed into the principal instrument of internal market management', but that it has also 'developed into a flexible instrument of commercial and external relations policy'. Often the extent and level of export subsidisation is unnecessary to maintain domestic prices. The *Kapitan Danilkin* affair is a good example of the subsidising of exports which the domestic market could easily have absorbed. Not only that, but the Union's tendency to over-subsidise exports is in itself a factor depressing world prices – thus increasing the level of subsidy which the EU has to pay to maintain the competitiveness of its exports on international markets.

> Thus the export subsidy system has a tendency to increase the level of Union prices in some markets and is one factor, along with the low value of the dollar, which helps to depress world prices in others, for example beef. The result is that the gap which the system is designed to bridge is widened; hence increasing the cost of disposals.
>
> (*The Management and Control of Export Refunds*, Court of Auditors Special Report, Luxemburg, 1990)

But the point of all this, too often pointless, activity is to support the incomes of farmers. This is something which it is clearly not doing – as the EU Commission itself admits:

> Income support, which depends almost exclusively on price guarantees, is largely proportionate to the volume of production and therefore concentrates the greater part of support on the largest and most intensive farms. So for example, 6 per cent of cereals farms account for 50 per cent of surface area in cereals and for 60 per cent of production; 15 per cent of dairy farms produce 50 per cent of the milk in the Union; 10 per cent of beef farms 80 per cent of the beef cattle. The effect of this is that 80 per cent of the support provided by the EU farm fund (EAGGF) is devoted to 20 per cent of farms which account also for the greater part of the land used in agriculture. The existing system does not take adequate account of the incomes of the vast majority of small and medium size family farms.[16]

The EU Commission's own analysis of the workings of the European Union's agriculture policies show quite clearly that the CAP has failed in what has always been regarded as the most important of the five main objectives of EU agriculture policy – to support farmers' incomes. These show that the gap between the incomes on large, efficient farms in the most prosperous areas and those on the small and under-capitalised holdings in the less favoured areas is continuing to widen.

The dairy sector is supremely typical of the Union's agricultural policy problem: it has many small farmers deemed needy of official support, while paradoxically having a few farmers with very large units, having a predominant

Relationship between EU12 agricultural income and agricultural support: increasing taxpayer expenditure fails to boost aggregate farm industry income

	1983	*1984*	*1985*	*1986*	*1987*	*1988*	*1989*	*1990*	*1991*	*1992*	*1993*
Real NVA at factor cost[1]	103.6	104.6	98.6	96.8	92.1	91.2	97.7	92.6	90.8	85.4	82.5
Total subsidy payments to agriculture	5511	6847	7464	7555	8621	9854	10891	11958	14389	16748	23847

[1]index – av. 1984–86 = 100
(NVA = net value added)
Source: *Eurostat Yearbook 1995*

Productivity raises individual incomes

	1983	*1984*	*1985*	*1986*	*1987*	*1988*	*1989*	*1990*	*1991*	*1992*	*1993*
Real net income/ worker	98.7	103	97.6	99.4	96.3	98.5	112.1	110.1	113.3	108.9	114
Agricultural population	11800	11250	10879	10108	9884	9428	8925	8599	8352	7619	7198

Source: *Eurostat Yearbook 1995*

SUBSIDIES AND INCOMES

MILLION ECU
TOTAL FARM INCOME
SUBSIDY PAYMENTS
GROSS REAL INCOME % OF 1984-6 AVERAGE

PRODUCTIVITY RAISES INDIVIDUAL INCOMES

Av. 1984-6 = 100
EMPLOYMENT IN AGRICULTURE '000S
INDEX : REAL NET INCOME/WORKER

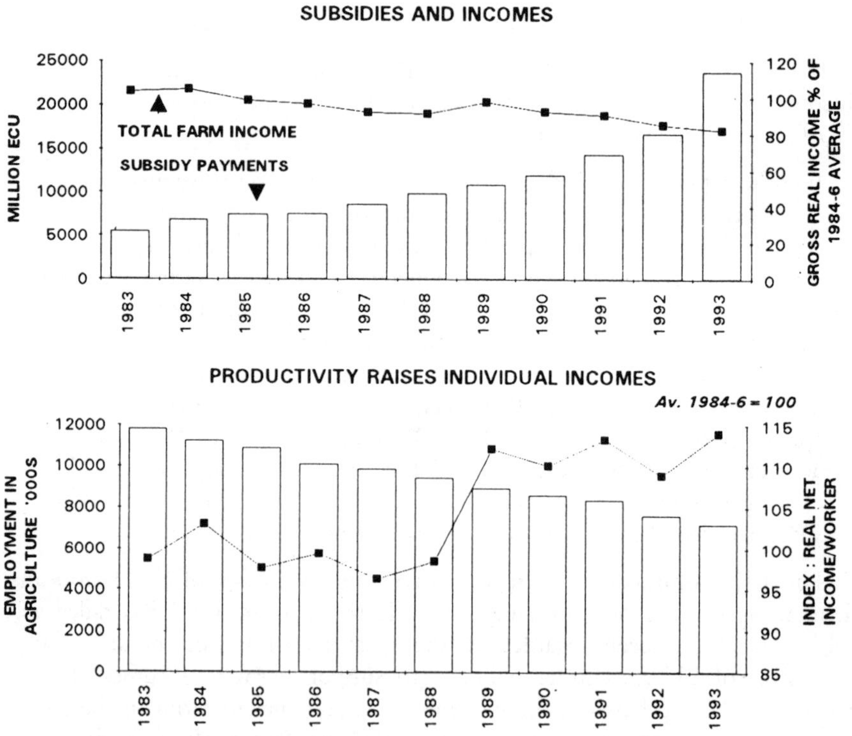

Figure 3.8 The futility of state support for agricultural markets: aggregate industry income falls as subventions increase (top bar chart); productivity increases individual incomes, whatever the level of support (bottom bar chart)

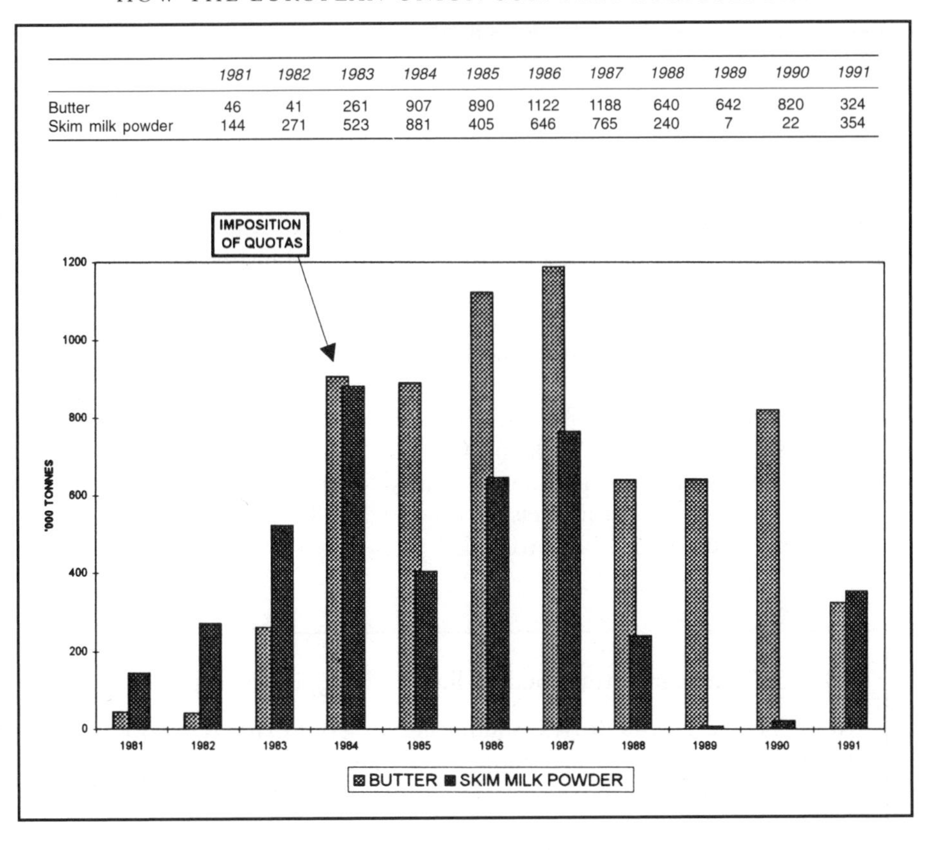

	1981	*1982*	*1983*	*1984*	*1985*	*1986*	*1987*	*1988*	*1989*	*1990*	*1991*
Butter	46	41	261	907	890	1122	1188	640	642	820	324
Skim milk powder	144	271	523	881	405	646	765	240	7	22	354

Figure 3.9 EU dairy product stocks ('000 tonnes)

share of production and with a high level of annual productivity increase. Where it is not typical of the EU market support regimes, is in having been subject through the 1980s to intensive political pressure to curtail its excesses; the dairy sector is the only one of the EU support schemes – until the May 1992 CAP reform – to have controls imposed on it during the last decade. It is also remarkable, despite these apparent controls, in having suffered very little reduction in output or diminution in the amount of taxpayer's money spent upon it.

Despite the imposition of production quotas in 1984, the EU dairy surplus has remained at high levels, never far from the 1984 level of production, and the cost of maintenance of the dairy policy has actually increased. The history of the dairy sector is thus also a warning to the politicians of how not to tackle the Union's strongly linked agricultural over-production and rural social problems.

Stabilise production the dairy modification policies of the 1980s may have done, but stabilise expenditure they did not and will not. Following the imposition of quotas in 1984, the bill for market support rose steadily until 1988, declined modestly with improvement on the international market in 1989, and in the 1990s is set to continue to increase, unless further restrictions are imposed on EU production.[17]

The problem is that the quota levels agreed in 1984 were far too large. They were set at the 1983 level of production which was then about 17 per cent above domestic consumption (some 30 per cent of which is in any case already heavily subsidised); the quota has been subsequently increased and consumption has declined. Consequently, the Community still in 1995 had between 13 and 14 million tonnes of raw milk equivalent which still has to be disposed of each year outside the Union market.

It is for this reason that dairy market support rose to an all-time high in 1988 of close to 6 billion ECU and why the average annual expenditure for the years since 1984 is over 5 billion ECU. And this did not include a special 'surplus disposal fund' used to dump some one and a third million tonnes of butter on international markets in 1986 and 1987. This operation alone cost the EU taxpayer 3.2 billion ECU.[18]

Box 3.1 The agricultural budget limit: The agricultural guideline

At the February 1988 Brussels summit meeting of EC heads of government it was laid down that future spending on agricultural support (EAGGF Guarantee) should not exceed the estimated 1988 level (27.5 billion ECU). In future years this limit was to be increased by 74 per cent of any increase in the Union's gross national product and also to be adjusted upwards by the average rate of EC inflation. The actual level of spending for comparison with this Guideline was to be defined as gross EAGGF Guarantee spending minus income from levies and expenditure on food aid. Limits were to be set on spending on the support of each of the main commodity sectors (the so-called commodity 'chapters' of the budget).

Table 3.2 EC agricultural support (billion ECU)

	1985	*1986*	*1987*	*1988*	*1989*	*1990*	*1991*	*1992*	*1993*
EAGGF Guarantee									
Gross	19.74	22.14	22.97	27.69	28.25	28.36	32.13	36.40	41.25
Net of levies	17.57	19.85	19.87	24.90	25.71	26.02	29.43	33.70	38.55
Agricultural Guideline				27.5	29.53	31.72	34.07	36.59	39.29
'Surplus'				2.6	3.82	5.70	4.64	2.88	0.75

Sources: Budget: up to 1990 – EC Commission; 1991–93 Agra Europe; Guideline 1989-93: Agra Europe

Union butter production has undoubtedly declined from its budget-bursting peak of very nearly 2.3 million tonnes in 1983 to 1.2 million tonnes in 1995. But production rises whenever the world market for dairy products falls – reflecting the extent to which the EU dairy industry relies on the butter intervention system to maintain its profits: if the world market is unprofitable, the intervention store remains the last resort of the industry's excess production.

During the later 1980s the Commission managed to obscure a 14 per-cent decline in butter consumption by increasing disposal under special subsidy schemes.[19]

THE CAP AND CONSUMERS

Consumers get a raw deal from the CAP. Because the farmer support system works essentially through the manipulation of markets – this is still largely so in the mid-1990s, even after CAP 'reform' and the GATT agreement – consumers pay for agriculture policy twice: once through higher food prices and again as taxpayers via the budgetary cost of operating the market management system.

A major component of the EU agricultural market support system is the 'import threshold'. Levies, customs duties and other taxes on imports maintain prices of imports for the major agricultural products at an average of 40+ per cent above world prices.[20] While it cannot be argued that the world price is a realistic price that consumers could expect to pay in a situation without an agriculture policy, it is obvious that a substantial proportion of this higher price is unnecessary and is therefore a tax paid by consumers. It involves a substantial transfer of income from consumers to taxpayers; an important proportion of this tax on consumers is in fact wasted and does not even benefit farmers. In addition to the cost to consumers of higher than necessary food prices, there is the cost of the operation of the domestic manipulation of the food market: the buying-up of surpluses, the subsidies to food processors, the disposal of surpluses and the very substantial cost of subsidies on the EU's exports of surplus products to make them competitive with the much lower world prices.

Various estimates of these combined costs of agricultural protection and support in the European agricultural common market have been made over the years. The main conclusion to be drawn from these studies[21] is that the 'hidden' cost of the CAP – the higher price paid by shoppers for food resulting from protection of the market – is always greater than the obvious, visible cost of the annual EU agricultural budget. It also increases in line with the rising budget cost of agriculture policy.

In the mid-1980s, when the budget cost of the CAP was in the region of 20 billion ECU, the additional consumer cost of higher food prices was estimated at more than 42.5 billion ECU.[22] Clearly, the larger the gap between the average world prices for major commodities and the internal

	1992	*1993*	*1994*	*1995*	*1996*	*1997*	*1998*	*1999*	*2000*
Cereal export restitutions	3.28	3.56	3.25	2.45	1.40	0.78	0.78	0.78	0.78
Compensation cost	0.00	5.00	13.61	14.65	16.20	16.20	16.20	16.20	16.20
Accompanying measures	0.00	0.18	0.54	0.80	1.40	1.80	1.85	1.90	1.95
Dairy	4.59	5.22	4.05	3.80	3.55	3.30	3.05	2.80	2.55
Beef	4.24	4.12	4.76	4.28	3.86	3.93	4.01	4.09	4.17
Rest	18.96	17.27	14.77	14.50	14.21	13.93	13.65	13.37	13.11
EAGGF guarantee estimates			40.98	40.48	40.62	39.94	39.54	39.15	38.76
EAGGF guarantee*	31.07	35.35	37.46						
Guideline	35.04	36.66	36.46	37.19	37.93	39.14	40.51	41.93	43.40

*Commission figures
Cereal export restitution expenditure: Agra Europe estimates

Figure 3.10 EU agricultural support spending in relation to spending ceiling (the 'Agricultural Guideline')

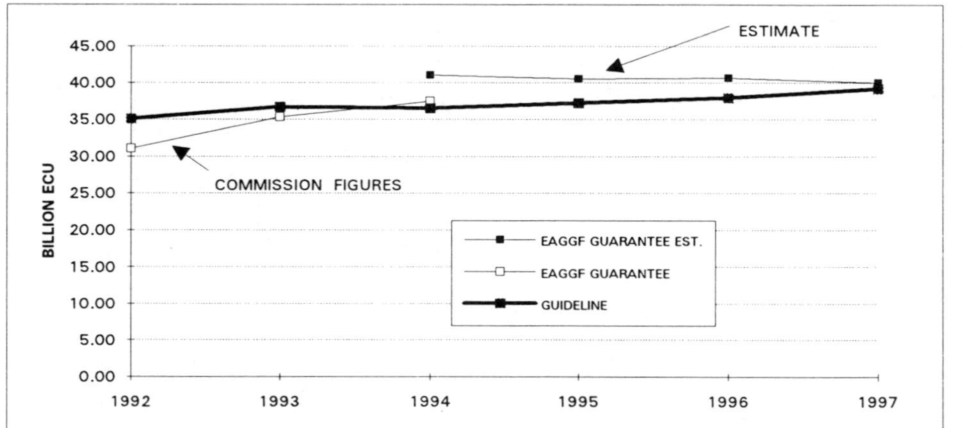

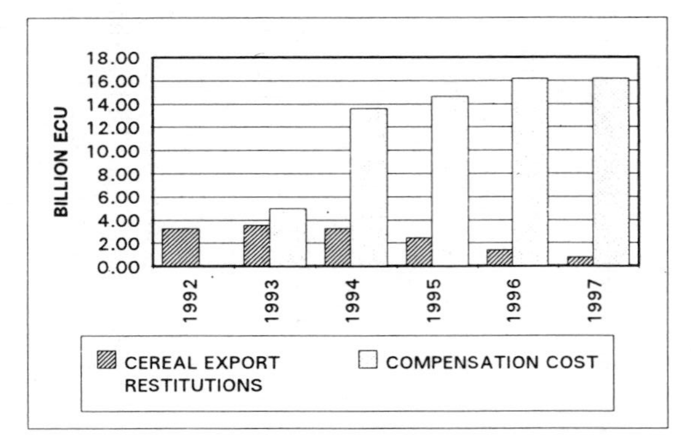

Figure 3.10 (continued)

CAP prices, the greater the consumer loss: when world prices are high the gap closes and the consumer loss diminishes; when they fall, the gap widens and the consumer cost increases. In theory, the 1992 CAP reforms and the 1993–94 GATT agreement should mean that the gap between EU prices and world prices should shrink and the consumer loss diminish.

The OECD estimates quoted earlier do not allow for the depressing impact of protectionist policies themselves on world prices, and thus its calculations represent the maximum loss resulting from the operation of the CAP. Complex computer modelling is needed to reach some estimate of what the cost of the CAP to consumers would really be if compared with the best alternative – a completely free EU and world market in food. Calculations by Tyers and Anderson[23] suggested that the overall impact of EU policies on the world market for the major commodities is to depress prices by an average 10 per cent. The impact is worse where there is a larger Union share of the international market. Because of the EU's dominant role in livestock production and exporting, its activities are calculated by Tyers and Anderson to depress the price of beef by 18 per cent and dairy products by 25 per cent.

When the impact of agriculture policies in all industrial market and less developed country economies is included in the calculation, then it is estimated that average world prices would be 9 per cent higher if these policies did not exist. It is therefore likely that the true 'tax on consumers' resulting from the CAP – and everyone else's protectionist policies – is not the 48 per cent of the OECD 1993 calculations, but approximately some 10 per cent less than the maximum OECD figure. It is however still excessive: Tyers and Anderson estimated that the cost of the CAP, taking into account the impact of non-EC protectionist policies, was in the 1980s in the region of $800 per non-farm household.[24]

Because of the wastage in the system, every dollar received by farm households costs the consumer/taxpayer $1.20. What is worse, however, is that the money is not filtering through to the 30 per cent of small and needy farmers for whom the policy is said to exist.

> In the case of the European Community, even if five or six of the $47 billion transferred to producers via the CAP reaches the target group of low-income farmers in the Community, this represents only one-tenth of the cost of CAP to consumers. In other words, consumers are forgoing $10 for every dollar received by needy farmers.

Put another way: 90 per cent of the taxpayer expenditure and consumer loss from prices that are too high are therefore unnecessary.

Defenders of the CAP and its high cost maintain that the EU is not alone in having high food prices and, since the cost of food now represents only 18 per cent of the average household budget, the cost to the individual consumer of the policy is relatively small. Consumer organisations strongly dispute this line of argument. They argue that groups within society have

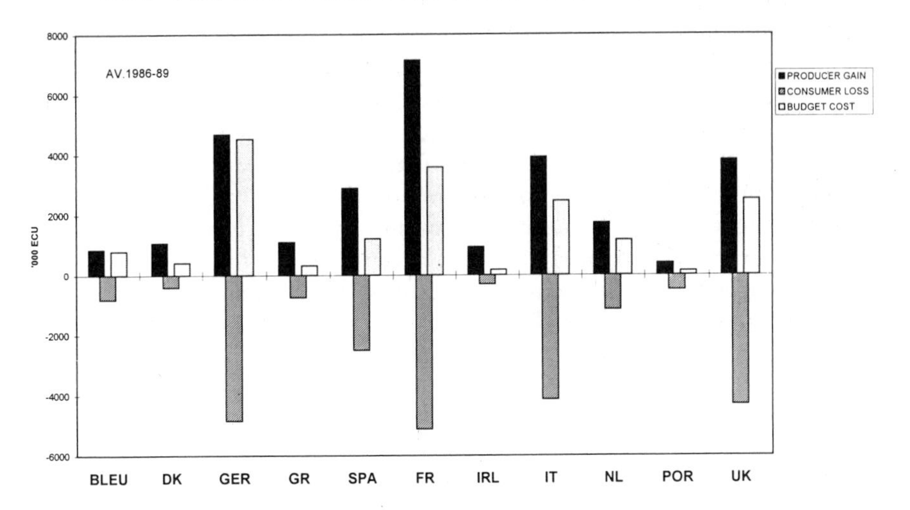

Figure 3.11 Gains and costs of the CAP

Source: Pros and Cons of Dismantling the CAP Research report, Arne Larsen, University of Copenhagen, 1994

less than average incomes and therefore food forms a much greater proportion of their total expenditure and thus the 30–40 per cent premium on EU food prices resulting from a regressive measure like the CAP is highly important for such groups. In the United Kingdom, for example, while the top 25 per cent of earners spend only 14 per cent of income on food, those in the lowest 25 per cent spend 29 per cent. The National Consumer Council, for example, argues:[25]

> There are large differences between the national average figures for the proportion of income spent on food from under 14 per cent for Germany to over 36 per cent for Greece. Also, within many member states there are variations in expenditure between regions. For example, in France the proportion of income spent on food varies from 16 per cent in Paris to 26 per cent in rural areas, while in Italy it varies from 27 per cent in the north east to 36 per cent in Campania, Sardinia and the south.

Even in the most prosperous economies there is, in any case, unjustified consumer loss. Money spent on food is not spent on other goods: the opportunity cost of the CAP therefore depresses the demand for other products and therefore the economy as a whole.

Consumer organisations also maintain that the CAP, while it may have stimulated the production of more than adequate supplies of basic foods, has reduced the choice of foods. This is because farmers are encouraged to produce 'standard qualities' which are easily officially recognised and can therefore be bought into the EU's surplus stockpiles. Food processors have found that because of this development and the limitations on purchase from outside the EU imposed by the import threshold system, traditional ingredients have become unavailable and the food industry has had to adapt its processes and recipes to suit the products which are available. One of the outstanding examples of this distortion is the supply of wheat to the British milling industry. Prior to UK membership of the EC in 1973, British millers bought hard wheat from North America and Australia to make the type of long-keeping bread popular with British consumers. The EU import levy system largely cut off this source of supply in the 1970s: a whole new and expensive process (the so-called Chorleywood baking process) had to be developed to allow unsuitable European soft wheat to be used in baking British-style bread.

Consumer organisations are also concerned about the impact of CAP market distortions on nutritional standards. A Danish consumer organisation study indicated quite clearly that the application of the CAP from 1973 had pressured consumers away from more expensive, quality food to less expensive, poor-quality food with a consequent decline in dietary standards.[26]

> A good example is the rise in cheese prices, which on average is about 2.5 per cent a year. We eat much less cheese than in 1966. On the contrary, we eat more than twice as much ice cream as in 1966. The price of ice cream has fallen about 3.4 per cent on average per year.

While cheese consumers are penalised by the EU import levy system and the milk powder intervention system, ice cream manufacturers benefit from butter-processing subsidies designed to get rid of the Union's internal milk fat surplus.

The problem for consumers is that the CAP is still – despite the reforms of the early 1990s, despite the GATT – a farm policy designed for the benefit of the agricultural industry. The achievement of the agricultural interest in preserving most of the major market-distorting features of the 'old CAP', albeit in different form, through the 1991–93 domestic policy reform process and through the continuing pressure for change brought about by the 1994 GATT agreement is testament to the skill and tenacity of the farm lobby in ensuring that its interests predominate in the farm policy formulation process. The objective of providing food supplies 'at reasonable prices' – to quote the Rome Treaty – takes a very second place in this process.

Though it is assumed that the CAP has been reformed and that the imposition of the GATT Uruguay Round conditions will force further reforms, the reality is that a great deal of the old policy still remains – most importantly, the commitment to support agricultural production at very high levels. As one OECD expert has pointed out:[27] 'Not only support for individual commodities, but support to the agriculture sector as a whole, can be maintained or increased by changing the form in which it is provided.'

Not only the European Union, but also the EFTA countries and the United States have managed to preserve their support of farmers by increasing direct payments to farmers. Preliminary assessment of post-1993 producer subsidy equivalents (PSEs) indicate that the levels for each country or group will not have changed despite recent claimed reforms.[28] What has happened is that the balance in payments to farmers has been shifted from direct market manipulation mechanisms to direct payments. The impact on production of these changes – in the absence of adequate production controls – is small. Efficient operators bolstered by state handouts will continue to produce.

4

THE AGRICULTURAL REVOLUTION

Peasants into tycoons

Down on the Dutch polders there is a farm where the cows are untouched by human hand. The cows come in from the lush pastureland of what has been for centuries one of Europe's great dairy regions, they enter their 'milking parlour' – a building more like a modern car factory than the rustic shippon of yore – machines, whirr, buzz and click, food drops into troughs in front of the cows' noses, mechanical hands and hoses wash the cows' udders and robot arms slap milking-machine cups onto the teats of each animal. Milking completed, in about half the time it used to take Gelda with her wooden pail, milking stool and practised hands to extract about a third as much milk – the details of the whole operation are recorded on the farm's comprehensive computer system. Not only are all the mechanical actions of the robotic cow herding system controlled by the computer, but so too are the details of each cow that passes through the system: her daily milk yield, food intake, weight gain or loss and other details that are essential to the precision husbandry that is rapidly replacing the picturesque but unprofitable farming beloved of the folklorists.

This Dutch dairy farm is of course at the frontiers of modern agricultural technology, and this degree of sophisticated use of computers and machinery to replace the routine work traditionally performed by men and women is unusual. Computerised information systems have, however, been used by farmers and the agricultural industry as long as in Europe's manufacturing industries. European agriculture now probably has more incentive to use machines in place of people; the post-war industrial revolution sucked people away from the countryside on an enormous scale and skilled wage rates in the areas of the most intensive farming close to the big urban areas soared. Farming can only be profitable, as in any other business, if output per person is maximised. For successful farmers, the computer is becoming the nerve centre which allows optimal exploitation of modern machines, crop chemicals and plant and animal genetic developments.

Farmers now use computers to predict the weather, to tell them when to spray their crops, to control breeding programmes and often to monitor the whole of their year's cultivating, planting, crop protection, harvesting and crop storage operations. The marketing of crops and livestock products is rapidly becoming highly computerised.

This widespread application of information technology is only a part of the technological revolution that has swept European agriculture in the decades since the Second World War, but it does illustrate that agriculture is as modern and technically advanced as most other industries. The owners and managers of the minority of farms which produce more than 80 per cent of the Union's agricultural output are likely to be as well-educated and as adept with modern management techniques as the managers of car or electronics factories.

Because, however, agriculture has been dependent upon the technological developments of other industries, of farm machinery manufacturers, of the electronics and computer industries and the chemical industry, its adoption and exploitation of what might be described as modern methods has been relatively delayed and at the same time telescoped into a relatively short period. Essentially agriculture has become a complex 'high-tech' industry in the three decades since 1960.

In those thirty years, in areas where large-scale modern agricultural techniques can be fully exploited, medium-sized farms have become large farms and small farms have been obliterated as the high profits to be earned from 'agribusiness' have encouraged the consolidation of small and medium-sized farms into large units. In the predominantly livestock farming areas, where large-scale crop farming cannot be practised, the technology has been concentrated more on stockbreeding than on mechanisation. New genetic strains have doubled milk yields, increased the growth rates of beef cattle, pigs and chickens, and reduced the costs of production, as meat has been produced faster, leaner and in greater quantity than ever before.

The combination of the rapid application of new techniques and the encouragement of lavish support under the CAP has increased productivity in agriculture by rates which far outstripped those in manufacturing industry. Between 1965 and 1980, productivity in agriculture in the six countries which originally formed the European Union increased by close to 6 per cent a year.[1]

In the 1950s continental European agriculture was generally regarded as primitive. Not only were human communities to a great extent still closed in rural areas, but this also applied to livestock and crop breeding; the average dairy farmer in most of western Europe would be working with cows which were largely the product of centuries of local breeding – too often local in-breeding. Average yields for dairy cows were less than 1,500 kg per year – that is, less than a quarter of today's average EU yield. Forty to fifty years ago there was little chance that the productivity of the average European cow could be increased by any significant amount – had the methods of breeding and husbandry practised at that time continued. But of course they did not. In the 1960s and 1970s European cattle breeding was revolutionised, principally by the introduction and widespread use of artificial insemination.

Suddenly, the average European dairy farmer was able to buy the services of the best bulls in the world. The rule of the local or village bull with very limited capacity for handing on high milk-yielding ability to its progeny was over. Increasingly, farmers were able to purchase semen from highly bred sires, produced from long lines of high-yielding cows. This was the major reason why milk yields trebled and production increased so rapidly in the European Community (as it was then) between the late 1960s and the mid-1980s when the lid was finally put on this productivity pressure cooker by the imposition of production quotas.

During that period, the dairy farming landscape of Europe changed dramatically. Gone were the fields full of different breeds, different colours and different styles of dairy cow. Within the average farmer's working lifetime the old breeds such as the Danish Red, the Saxon, the Dairy Shorthorn were almost eliminated from Europe's pastures to be replaced by the uniform black and white of the Friesian and the Holstein.

During the period 1970–83, the numbers of animals from indigenous breeds of cattle in milking herds decreased by an average of 50 per cent in all of the original six member states of the European Community.[2]

As a result of the straightforward improvement in the genetic quality of cows – and without such productivity boosters as BST – output per cow in the EC10 has increased by more than 25 per cent during the twelve years to 1992; average productivity is continuing to increase, as the 5000-litre-plus cow becomes the norm rather than the exception and the 8000 litre a year 'super cow' becomes more common. In the last three decades productivity in the EU dairy farming industry has increased by more than 50 per cent.

This has been the major thrust behind the growth of the European Community's massive dairy surplus in the late 1970s and early 1980s. By 1984, when an exasperated Council of Ministers finally clamped down with a quota system which attempted to pin production down to the 1983 level, the Union was producing approximately 18 million tonnes of milk surplus to its own consumption. The EC's surplus alone was equal to the entire output of the New Zealand dairy industry. The transformation of the European Community from a dairy-importing region into the world's major exporter between 1969 and 1983 was achieved with a smaller herd of cows and fewer dairy farmers than existed in the late 1960s. The quality of both had increased mightily in the intervening fifteen years.

What EU dairy experts are only too well aware of is that the removal of quotas would immediately stop the decline in the dairy herd, would probably lead to a modest increase in cow numbers and, within the two and a half years it takes a heifer calf to become a milker, the Union would be back in the state of increasing over-production, accumulating surplus and growing expenditure which it faced in the early 1990s. The stabilisation of milk deliveries and a growth in excess processing capacity has also speeded up the process of rationalisation of the milk-processing industry.

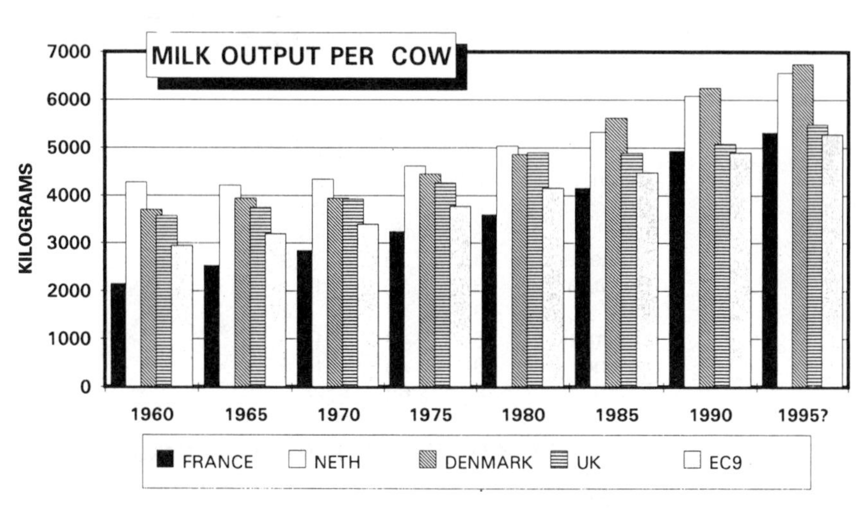

Figure 4.1 Milk yields in major EU dairying countries

According to the EC Commission's figures,[3] the number of dairy plants in EC9 declined by 11 per cent between 1982 and 1985; the problem of under-utilisation of plant created by dairy quotas is likely to have led to further contraction of the industry and to rationalisation of the remaining plants into much larger units.

The most recent detailed survey of the EU dairy industry confirmed the developing domination of the EU industry by the larger-scale producers.[4] Whereas farmers with fifty cows or more represent only 6 per cent of the Union's 1.64 million dairy farms, they own almost a third of the total 25 million cow Union herd. These large producers have an average herd size six times greater than the overall EU average of 13.6 cows. The concentration of large herds varies widely from country to country. While more than 56 per cent of Dutch dairy farms have an average of seventy-seven cows, only 16 per cent of the French herd is in the large category. The actual number of farms with large herds in France is, however, about the same as in the Netherlands.

The survey showed that the large herds are concentrated in the areas with greatest geographical and climatic advantages for fodder production or for access to cheap, imported feed supplies. The large herd category of farmers tend to concentrate more animals on fewer hectares: while owning over 30 per cent of the Union herd, the large producers utilise only just over 20 per cent of the total Union forage area devoted to dairy production.

The changes were almost as radical in crop farming as in livestock. During the thirty years to 1985, the labour force in the original EC Six declined by more than 60 per cent – more than 12 million people. At the same time, the numbers of tractors on farms in the EC Six increased by over 70 per

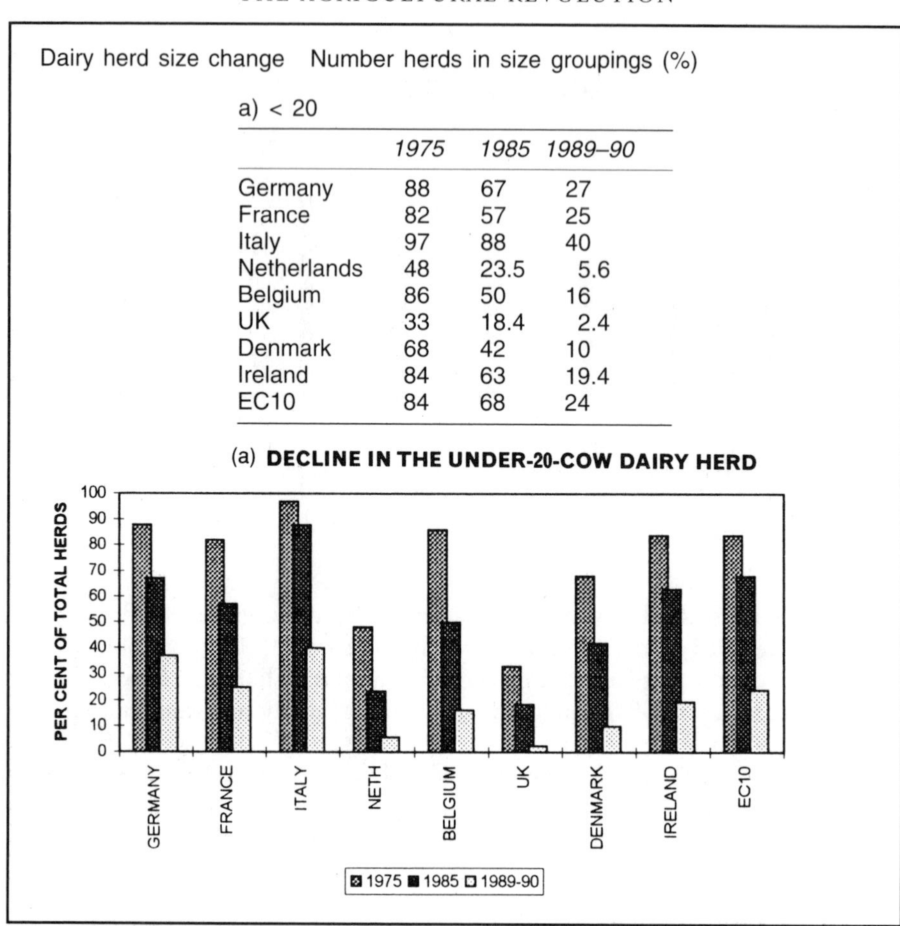

Dairy herd size change   Number herds in size groupings (%)

a) < 20

	1975	1985	1989–90
Germany	88	67	27
France	82	57	25
Italy	97	88	40
Netherlands	48	23.5	5.6
Belgium	86	50	16
UK	33	18.4	2.4
Denmark	68	42	10
Ireland	84	63	19.4
EC10	84	68	24

Figure 4.2 Change in the structure of the EU dairy farming industry

cent as machines replaced human and animal muscle power. During this same period average yields of the major crops increased by startling amounts; the word 'startling' is used advisedly, since undoubtedly the increases in crop production did startle the EU politicians and threw all their plans for the smooth transition from peasant agriculture to what they expected would be a pattern of cosy family farms into complete disarray.

Wheat yields, for example increased by over 60 per cent in the period 1965–1980.[5] Similar patterns were shown in the other major crops – feed grains, sugar beet, potatoes; and of course at the same time these rises in crop productivity were matched by increases in farm size.

These increases in productivity have of course not been achieved without considerable cost – particularly to the taxpayers of the countries concerned. During the 1970s and 1980s investment in agricultural research increased

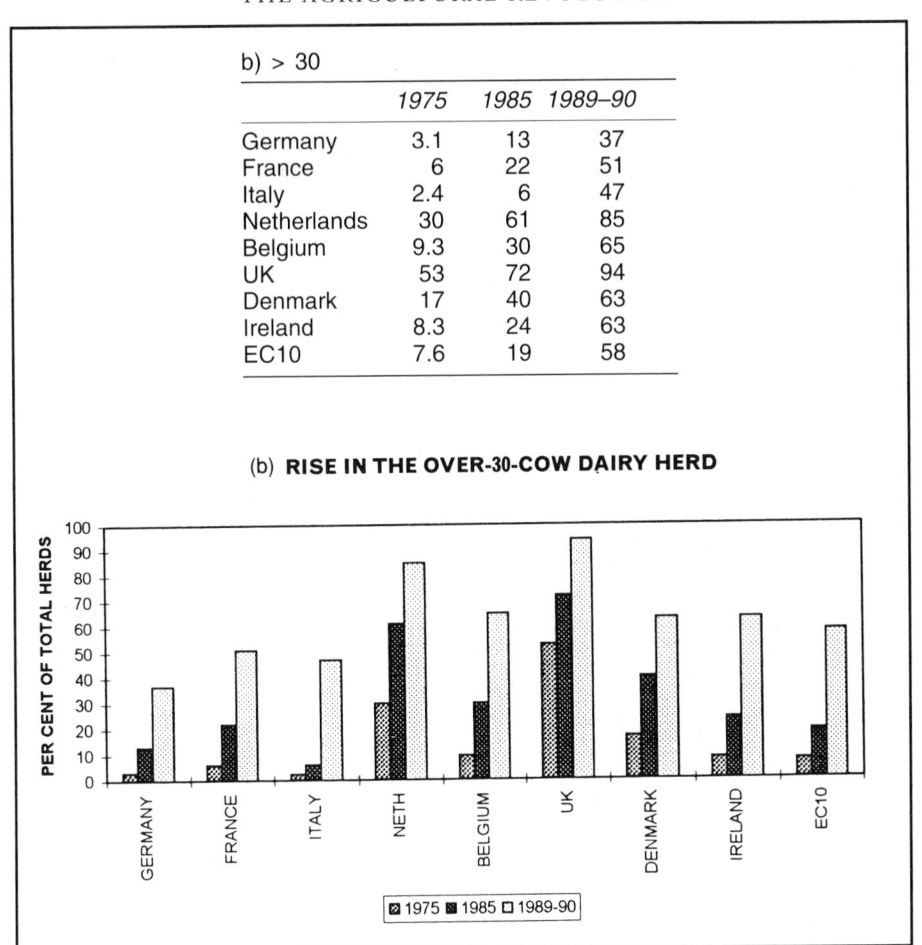

b) > 30

	1975	1985	1989–90
Germany	3.1	13	37
France	6	22	51
Italy	2.4	6	47
Netherlands	30	61	85
Belgium	9.3	30	65
UK	53	72	94
Denmark	17	40	63
Ireland	8.3	24	63
EC10	7.6	19	58

Figure 4.2 (continued)

dramatically in most developed countries but particularly in the European Union. In the Netherlands,, for example, the national research investment per farm increased from $398 to $1563[6] in 1980; in Germany, investment in research increased from $145 per farm to close to $400 per farm in 1980. In the United Kingdom, the increase between 1970 and 1980 was from $382 to $964 per farm. In international terms, only one country outside the European Community exceeded the EC level of investments in agricultural research, and that was Australia, where research investment per farm increased from $616 in 1970 to $1130 in 1980; even the United States only increased from $247 to $537 per farm in the 1970–80 period.[7]

There was also a substantial investment in agricultural education in the post-war decades, with the numbers of agricultural college graduates per 10,000

Table 4.1 Indexes and average annual growth rates of agricultural total factor productivity for selected countries by stage of development, 1970 and 1980

Country	*Total factor productivity level* 1970*	1980** *Index*	*Average annual growth rate, 1970–80* %
Australia	61.67	79.83	2.61
Austria	51.65	63.57	2.10
Belgium	83.25	106.48	2.49
Canada	70.53	82.24	1.55
France	59.97	73.20	2.01
West Germany	63.21	74.88	1.71
Netherlands	69.91	95.94	3.22
New Zealand	89.77	95.80	0.65
Sweden	64.33	75.72	1.64
United Kingdom	63.59	73.74	1.49
United States	75.27	100.00	2.88
Ireland	34.26	43.95	2.52

*1967–72 averages
**1975–80 averages

Source: US Department of Agriculture, Agricultural Economic Report No. 644, 1991

Table 4.2 Agricultural research and education for selected countries by stage of development, 1970 and 1980

	Research investment per farm		*Agricultural college graduates per 10,000 farmworkers*	
Country	*1970*	*1980*	*1970*	*1980*
	$*		No.	
Austria	26.33	51.20	5.79	12.41
Belgium	97.66	317.59	10.75	54.11
France	42.86	171.57	2.11	4.81
West Germany	144.69	397.11	7.39	31.98
Netherlands	397.92	1563.37	10.31	42.05
New Zealand	474.26	845.99	16.28	58.67
Sweden	188.45	623.72	6.55	34.02
United Kingdom	382.35	963.85	7.49	35.60
United States	246.75	536.97	21.21	135.14
Ireland	71.53	103.10	3.91	11.06

**In constant 1980 dollars*

Source: US Department of Agriculture, Agricultural Economic Report No. 644, 1991

farmworkers increasing by very significant numbers in the 1960s and 1970s. In West Germany, for example, the number of agricultural graduates per 10,000 workers increased from 7.4 in 1970 to 32 in 1980. In the Netherlands the increase was from 10.3 to 42, the United Kingdom from 7.5 to 35.6.[8]

Of course the increases in productivity were not limited to the European Community (European Union). According to a study by the US Department of Agriculture, world-wide agricultural production has increased by two-thirds since the mid-1960s, principally due to improvements in the quality of plant varieties and livestock.[9] Cereal yields increased 65 per cent in the period between the mid-1960s and the late 1980s, with an overall increase in world production of 60 per cent. The Department lists the most important technological developments responsible for these substantial increases in productivity and production as the use of hybrid seeds, more efficient use of fertilisers and the greater use of pesticides, the use of more machinery, and the development of disease control in plants and livestock. The investment in research, education and agricultural advice and extension services are also acknowledged as being major contributors to the improvement in agricultural productivity.

Agricultural productivity in western Europe can be said to have gone through two main phases during the period since the Second World War: a 'development' stage where many peasant farms disappeared, farm size increased and modern methods of livestock and animal husbandry were widely adopted, and the later, current stage, where the latest techniques are being increasingly adopted by the most advanced farmers who now control 60 per cent or more of the Union's farmed land area. The effects of the first stage was to increase aggregate production by dramatic amounts: more than 100 per cent in the case of cereals and by similar amounts in the dairy sector – in a period of little more than fifteen years.

The second, modern phase, has produced and is producing a steady and sustained increase in productivity in both the arable and livestock sectors. This is the result of the application of the most sophisticated innovations being provided by the plant and animal geneticists, chemists and engineers to the top level of most advanced farmers. In crop farming it can be expected that the uptake of modern varieties, in particular, will continue to maintain annual increases in productivity at between 1.5 and 2.5 per cent a year into the late 1990s. Plant breeders are confident that the evolution and application of new hybrid varieties could raise annual increases in prime farming areas to 3.5 per cent by the late 1990s and sustain such an increase into the second decade of the next century.

Such a sustained increase in output would, however, depend on the continued use of a high level of fertilisers and pesticides. It cannot, though, be assumed that current high levels of both fertiliser and pesticide will be able to be sustained in the face of the growing pressure in most EU countries for legislation restricting the use of such materials (Chapter 8). It can be

Table 4.3 Cereal yields (100 kg/ha) in member states, 1976 and 1990

	All cereals			*Soft wheat*			*Barley*			*Maize*	
	1976	*1990*		*1976*	*1990*		*1976*	*1990*		*1976*	*1990*
NL	46.7	69.0	NL	52.0	76.3	IRL	42.4	56.8	I	60.7	76.0
IRL	41.9	62.1	DK	50.0	72.2	NL	42.2	53.7	B	54.2	76.3
UK	39.7	61.7	IRL	44.9	81.0	B	41.0	60.7	D	53.4	71.0
B	39.7	63.1	D	43.7	66.7	D	39.7	55.9	F	44.8	66.9
D	39.1	58.9	UK	43.6	69.9	UK	38.4	51.9	GR	39.4	100.8
F	37.3	62.4	B	41.6	65.7	DK	36.1	52.6	E	38.5	64.7
DK	36.6	58.5	F	40.1	66.3	F	33.2	57.1	P	14.5	30.8
I	34.1	40.2	I	30.3	40.6	L	26.2	45.7	EUR 12	45.4	68.8
GR	23.7	37.5	L	24.4	51.0	I	25.1	36.5	EUR 10	50.1	72.1
L	23.6	45.9	GR	23.6	27.2	GR	22.0	23.8			
E	18.9	24.8	E	15.8	23.5	E	19.2	21.5			
P	11.4	19.6	P	11.5	17.0	P	7.6	11.9			
EUR 12	32.1	48.2	EUR 12	32.9	56.9	EUR 12	31.4	40.5			
EUR 10	36.5	56.2	EUR 10	38.4	63.8	EUR 10	35.9	52.8			

Table 4.4 Production of cereals in 1990 and cumulative growth in production by member state, 1976–90

Member state ranked in order of quantity produced	*Cereal production 1990 million tonnes*	*Share of EUR 12 production %*	*Cumulative growth in production 1976–90 %*
France	57	35.1	60
Germany	26	16.2	28
UK	23	13.8	55
Spain	19	11.4	41
Italy	17	10.2	11
All others	22	13.3	39
EUR 12	164	100.0	42

expected that restrictions on nitrate and other fertiliser use could begin to limit output by the end of the decade. Limits on pesticide use may, however, take longer to be applied. But there is the possibility, indeed the probability, that biotechnology could come to the rescue: engineering a plant's genes make it possible to employ more environmentally acceptable methods both of maintaining soil fertility and keeping pests and disease at bay (Chapter 9). Most notably, wheat and other cereals can be bred to produce higher yields with lower inputs of nitrogen and other fertilisers.

Clearly, the effect of the increases in productivity is not spread either evenly throughout the industry or among the countries of the Union. The larger, more geographically advantaged and most efficient sector of the industry is likely to be able to achieve at least an annual increase in productivity of 2.5 per cent and could well exceed 3.5 per cent. Marginal producers are going, and will continue to go, to the wall. Much of the grassland that was ploughed up to grow cereals during the EC-financed cereal bonanza of the 1975–82 period will go back into grassland and create problems in the beef and sheep sectors in the late 1990s. Production will continue to increase on the best land.

The important question is therefore: will the increase in production from those who remain in the industry outweigh the decrease in output from those who give up farming? Let us take cereal production as a specific example. In order to arrive at some approximation to the likely trend of development it is necessary to make a number of assumptions based on the likely practical details of EC cereal production in the 1990s.

First, it is likely that the number of cereal producers who give up growing grain in the next ten years (that is, to 2005) will be in the five main cereal growing countries of the Union – France, Germany, the United Kingdom, Spain and Italy. Cereal production in the other member states has close relationships with livestock and other specialist production such as malting and durum wheat-flour production, which make them less susceptible to the main thrust of EU cereal production cost and return developments. It is known that more than 80 per cent of cereal production in the major cereal-growing countries is in the hands of the minority of large and relatively efficient cereal growers.

Second, after allowing for specialist producer and livestock linkage factors, it is probable that farmers holding about 12 per cent of production in the five countries listed above can be classified as marginal cereal growers; these people are likely to be squeezed out of production by the EU's current pricing policies. Third, the remaining cereal growers will increase their productivity over the period 1989–95 by an average of somewhere between 1.5 and 3.5 per cent a year.

These productivity developments are in part a result of changes in the EC farm structure, but are themselves also the main stimulation for the radical alteration of the Union's farm size, labour force and degree of specialisation.

The degree of intensification of agriculture in the post-war era can be judged from the changes in the major indicators: reduction in agricultural working population, changes in holding structure (most importantly the increase in size), increased mechanisation, fertiliser and pesticide use and increasing output of crops and animal products. In the 1940s more than a quarter of the population of northern countries of what is now the EC12 worked on the land; today the total is less than 7 per cent, and probably more than half of those who remain on the land also only work part-time, with the main part of their income arising from non-agricultural employment.

Agricultural output in the same forty-year period has more than doubled. The degree of rural depopulation and agricultural intensification of course varies from country to country; the effect has been most marked in areas such as northern France, northern Italy and parts of Germany where areas of essentially peasant farming have, within a generation, changed to large-scale, intensive agribusiness.

The most significant statistic among the mass inevitably produced on farm structure is that today in the EU12 the 6.2 per cent of farm holdings greater than 125 acres (50 hectares) hold more than 50 per cent of the total farmed area.[10] In the 1950s the proportion held by the larger farmers was less than 20 per cent. In intensively farmed countries like the Netherlands and the United Kingdom the numbers of tractors very nearly doubled in the 1950–85 period; the increases were of course greater in those countries which were more agriculturally backward after the Second World War. In Italy, the number of tractors increased more than tenfold. Hedge and woodland removal and both major drainage and under-drainage schemes were widely extended in the three and half decades to 1985: by the early 1980s more than 50 per cent of the land area of the north-western European countries had been under-drained.

The production of all the major products increased vastly as a result of these productivity increases. In the major cereal-producing countries such as

France the average yield increased by close to 80 per cent between 1970 and 1990, from 3.4 million tonnes per hectare to close to 7 tonnes/hectare, sugar yields increased by 41 per cent from 6.7 tonnes a hectare to 9.5 tonnes a hectare, while milk yields in the main dairy-farming countries of the EU such as the Netherlands increased by 39 per cent, from around 4000 to close to 6000 litres per cow.

The importance of these figures lies not in the increase itself but – in the context of the type of policy operated by the EU – in the cost of the excess in terms of taxpayer expenditure on subsidies to reduce expensive EU-produced commodities to the world price level. If we take beef as an example: 8 per cent over-supply does not appear to be excessive – until it is realised that this excess represents 500,000 tonnes a year, each tonne of which costs an average 750 ECU in export subsidy to dump on the world market. Put another way, it costs close to 4 billion ECU to dispose of this 'small' excess over domestic consumption.

Structural change in the agricultural industry has been one of the factors which have stimulated the migration of people from the countryside to the towns. In France, for example, the farm population fell 59 per cent between 1960 and 1990, from 3.43 million people to only 1.39 million in 1990. In the original six countries of the EC, the farm population fell 54 per cent from 10.4 million in 1960 to 4.8 million in 1990. At the same time, farms have become fewer and inevitably larger: the number of farms in the original EC6 has fallen 20 per cent from 5.88 million to 4.7 million during the 1970s and 1980s.

There has also been a considerable element of 'catching up' by the originally less efficient countries with the most efficient, over what might be called these last two 'modernisation decades', 1970–90. While there have been dramatic reductions in the numbers of people employed on the land in France and Germany, in other countries where there had already been significant modernisation in the immediate post-war decades, such as Denmark, Sweden, the Netherlands and the United Kingdom, there was much less change in the structure of agriculture. In these latter countries there was much less scope for shedding labour from agriculture, less to gain from farm enlargement and less to be gained in terms of increased farm incomes. The farmers of Denmark, the Netherlands and the United Kingdom were none the less substantial contributors to the lavish subsidised production encouraged by the excessive support of the CAP.

The impact of farm modernisation under the aegis of the CAP was most marked in those large agricultural countries of western Europe where the 'peasant' farmer had hung on long beyond his obsolescence in the northern and western fringes of the Union: France and Germany.

The number of people employed on farms in Germany fell in the 1970s and 1980s by more than 30 per cent between 1975 and 1990, from 2.2 million to 1.5 million; in France, the reduction was similar: by close to 34

Utilised agricultural area by size classes of holdings (1000 ha)

	1975	1989/90	1975	1989/90	1975	1989/90	1975	1989/90	1975	1989/90	1975	1989/90
	less than 5 ha		*5–20 ha*		*20–50 ha*		*50–100 ha*		*more than 100 ha*		*Total*	
EUR 12		8333.6		19335.7		28131.4		22303.2		36556.4		114660.3
Germany	759.2	486.0	4310.7	2751.4	5258.7	4960.0	1456.2	2553.4	613.8	1017.9	11639.4	11768.7
France	705.3	516.2	5332.5	3254.1	11200.1	9346.7	7169.1	8736.8	5056.6	6682.6	28758.3	28536.4
United Kingdom	99.1	76.2	903.2	765.0	2567.2	2001.0	3200.3	3008.0	9709.2	10647.8	16379.9	16498.0

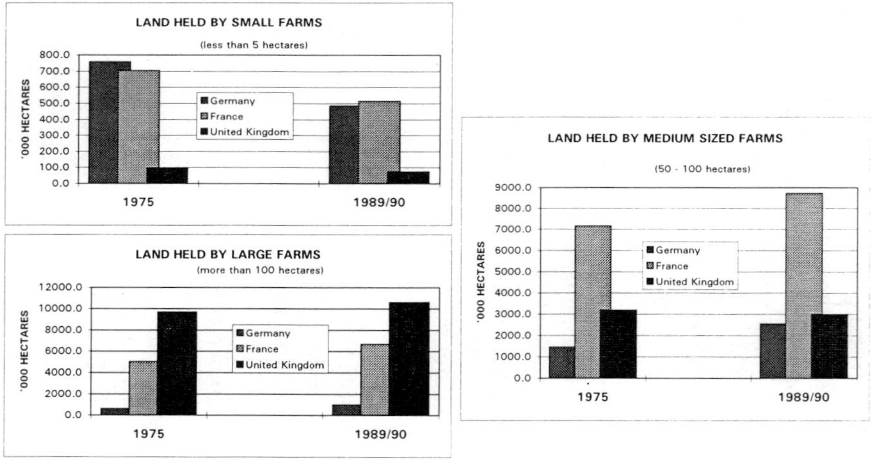

Figure 4.3 Change in structure of farms by size

Distribution of holdings by size classes of standard gross margins (1000)

	1975	1989/90	1975	1989/90	1975	1989/90	1975	1989/90	1975	1989/90	1975	1989/90
	less than 6 ESU		6–12 ESU		12–40 ESU		40–100 ESU		more than 100 ESU		Total	
EUR 12		5406.8		996.2		1242.6		421.2		102.5		8169.3
Germany	506.6	265.7	198.8	104.3	189.1	220.5	11.5	67.7	1.8	7.0	907.8	665.1
France	666.6	364.2	322.6	145.7	289.6	362.5	30.6	122.1	4.9	22.5	1314.3	1017.0
United Kingdom	1222.4	92.5	51.2	27.5	72.3	58.6	18.0	44.5	5.1	22.0	269.0	243.1

ESU: European size unit = each 1000 ECU of annual income

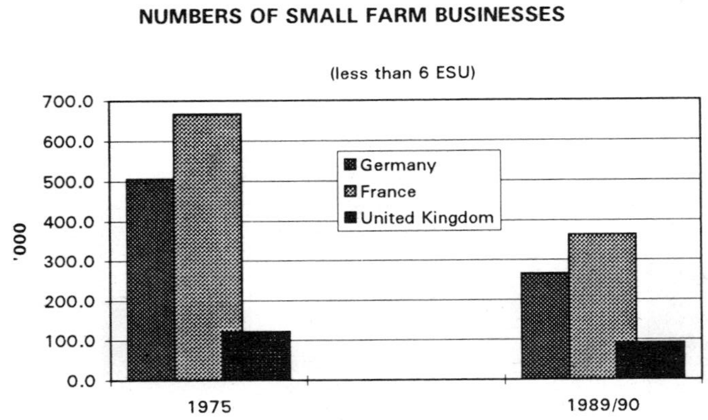

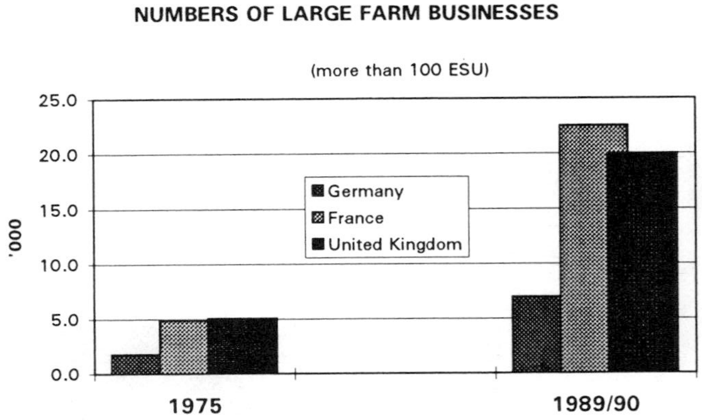

Figure 4.4 Change in structure of farms by size of business

Labour force: in annual work units (1 annual work unit = 1 full-time worker)

	1975	1989/90	1975	1989/90	1975	1989/90
	Total annual work units (1000)		by 100 ha		by holding	
			utilised agriculture area			
EUR12		8009.5		000		1.0
Germany	1233.6	786.8	9.9	6.7	1.4	1.2
France	1949.7	1389.3	6.6	4.8	1.5	1.4
UK	625.7	473.5	3.8	2.9	2.2	1.9

TOTAL ANNUAL WORK UNITS

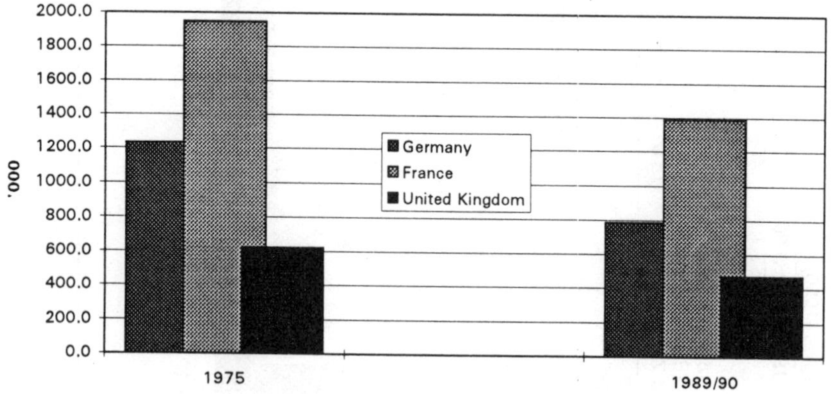

WORK UNITS / 100 HA

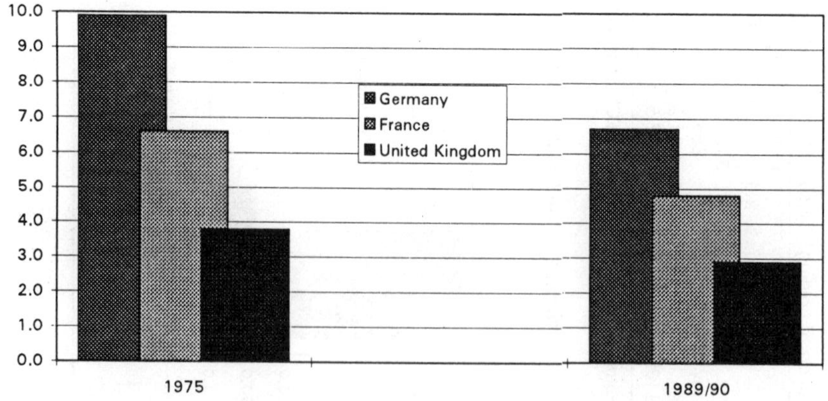

Figure 4.5 Change in labour force in relation to land area

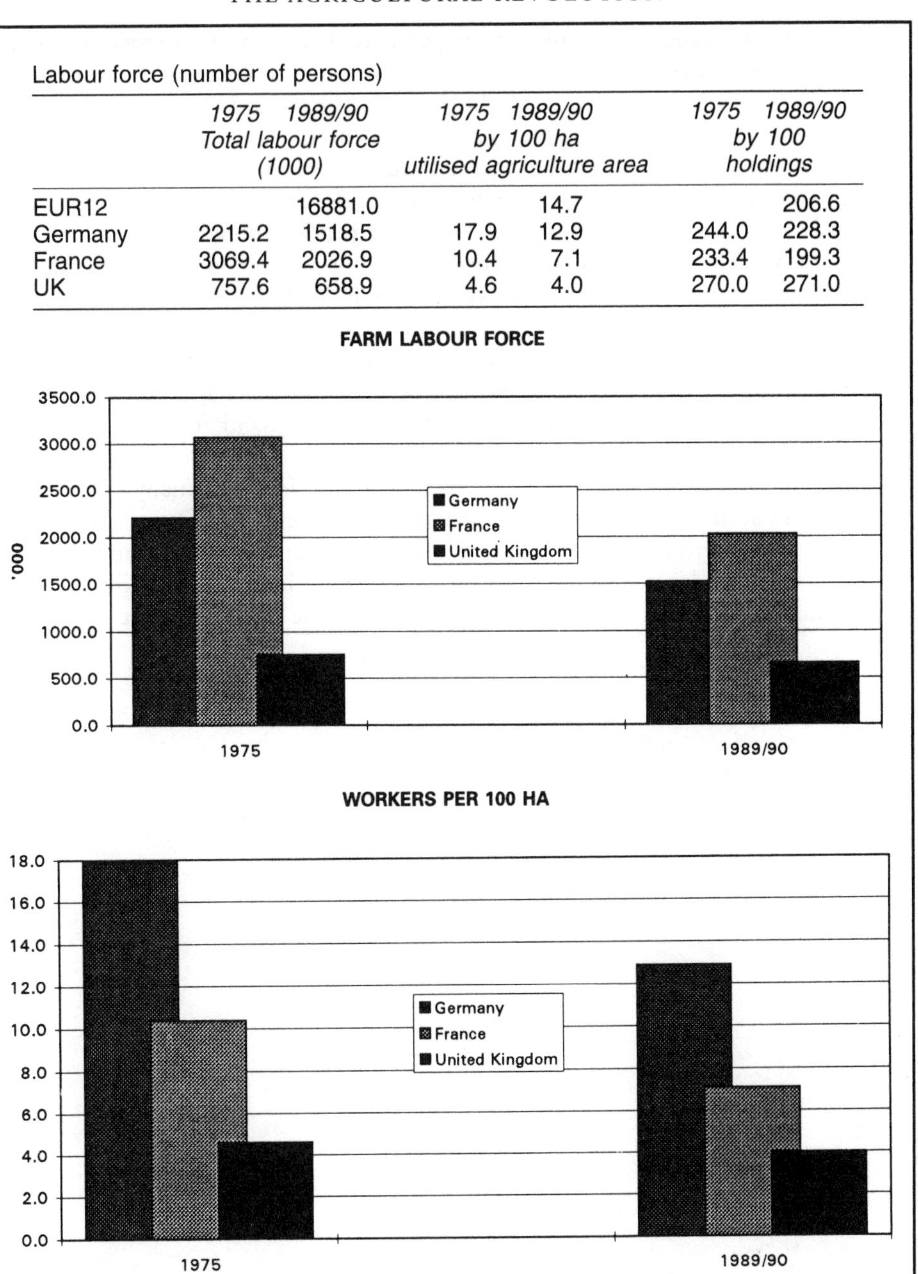

Labour force (number of persons)

	Total labour force (1000) 1975	Total labour force (1000) 1989/90	by 100 ha utilised agriculture area 1975	by 100 ha utilised agriculture area 1989/90	by 100 holdings 1975	by 100 holdings 1989/90
EUR12		16881.0		14.7		206.6
Germany	2215.2	1518.5	17.9	12.9	244.0	228.3
France	3069.4	2026.9	10.4	7.1	233.4	199.3
UK	757.6	658.9	4.6	4.0	270.0	271.0

Figure 4.5 (continued)

per cent, from just over 3 million to just over 2 million. This reduction in the number of people on the land inevitably led to a reduction in the concentration of workers in relation to the area of farmed land. Not surprisingly, the greatest reduction was in Germany, with a fall in workers per 100 hectares from nearly eighteen to just under thirteen, a reduction of nearly 28 per cent. Proportionally, however, France has made the largest reduction, by nearly 32 per cent from 10.4 to 7.1 workers per 100 hectares.[11]

Possibly a more accurate measure of the improvement in labour productivity which has taken place in the European Community/Union in the 1975–90 period are number of full time working hours (in EU terminology 'work units' = 1800 hours per year) devoted to each hectare of farmed land. According to the EU statistics the number of work units per 100 hectares fell from just under 10 to just over 6.5 in Germany and from 6.6 to 4.8 per cent in France. Significantly, though the crude numbers of people on the land and working on farms have changed little in the 'initially efficient' countries such as the United Kingdom, the Netherlands and Denmark, even in these countries there has been a marked reduction in the number of working hours devoted to each 100 hectares of land. In the United Kingdom, the number of work units/100 hectares fell by close to 14 per cent.

PART II

EUROPEAN AGRICULTURE IN THE WORLD ECONOMY

5

GOING TO THE TRADE WARS

Over-production, surpluses and international politics

New Zealand farmers get 5 cents (US) a litre for their milk; EU farmers get 30. This is why the NZ dairy industry can ship butter 13,000 miles round the world to Europe and still undercut British farmers by 25 per cent in their own market – or they would do if Brussels didn't slap a 25 per-cent tax on the imports to stop them benefiting UK consumers in this way.

Though the EU levy system which operated until 1995 has been replaced by a variable tariff (as a result of the 1994 GATT agreement) which will gradually be reduced, it is likely to have broadly the same effect as the variable levy system of maintaining a substantial difference between EU internal prices and the world price of the major agricultural commodities well into the first decade of the next century. The GATT tariff and tariff reduction process agreed in 1993 is a compromise which effectively allows the EU to continue to discriminate against imports and maintain much higher prices for food within the Union than those generally paid outside.

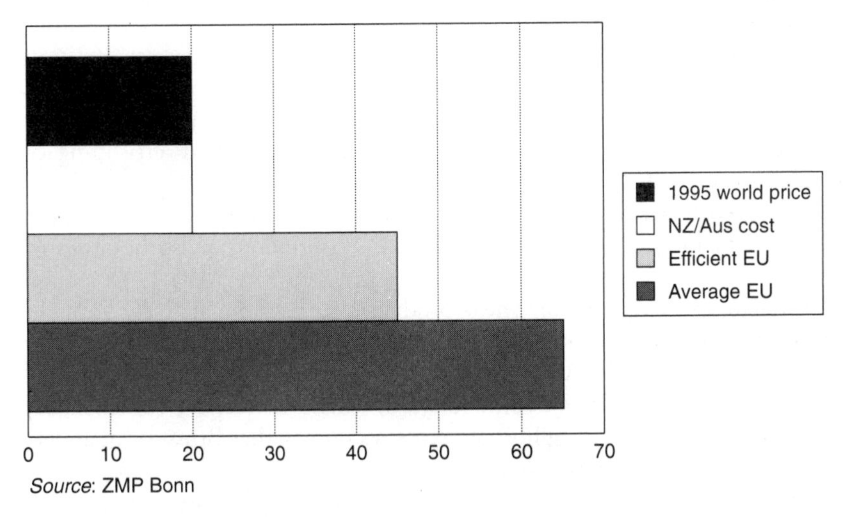

Source: ZMP Bonn

Figure 5.1 Cost of milk production

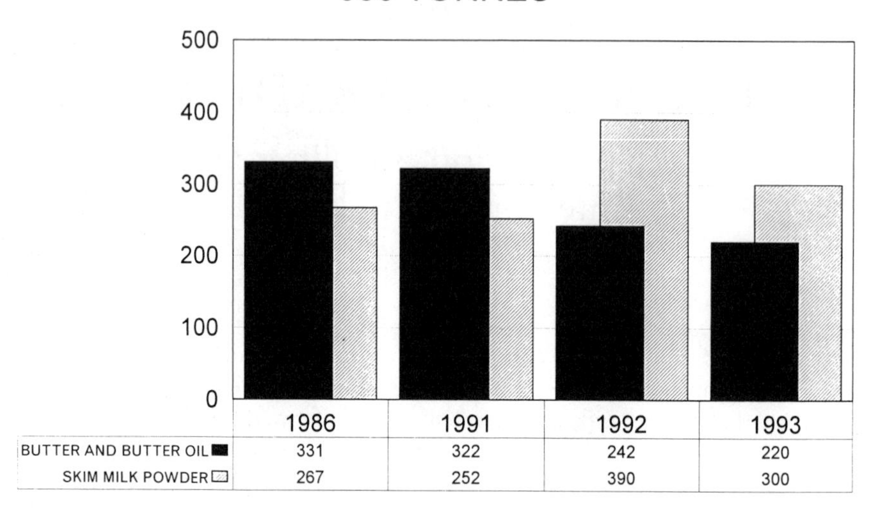

	1986	1991	1992	1993
BUTTER AND BUTTER OIL	331	322	242	220
SKIM MILK POWDER	267	252	390	300

Figure 5.2 EU dairy exports

The iniquity of the exclusion of efficiently produced and therefore cheaper food from the EU market by the Union's largely unbroken wall of agricultural protection is made worse by the Union's internationally unacceptable habit of dumping the surpluses which it creates on the international market.[1] To do this it pays massive subsidies – usually equal to half the value of the product – in third-country, non-EU markets where New Zealand and other efficient, unprotected producers have to sell their produce squeezed out of the European market.

NZ butter is an extreme example. Similar situations, however, exist in the wheat, beef, poultry-meat and milk-powder markets: not only are world producers barred from the European market, but they have to suffer destruction of non-European markets by the EU's subsidised jettisoning of its surpluses as well.

The United States also subsidises its farmers – but claims that its subsidies are needed mainly to counterbalance the EU's continuous lavish bounties to farm exporters.

Australian and Argentinian wheat producers suffer from it, New Zealand and South American sheep producers suffer from it, Australian and NZ beef producers suffer from it. American wheat growers would also suffer from it – if their government did not come forward, whenever necessary, with export subsidies to match the $15 billion of taxpayer's money that Brussels squanders on subsidising farm exports each year.

This subsidising activity is at the roots of the running diplomatic battle that has gone on between Brussels and Washington and between the Union and

World dairy exports ('000 tonne milk equivalent)

	1985	*1986*	*1987*	*1988*	*1989*	*1990*
EU	12240	11230	13834	16502	14226	12878
New Zealand	4001	4216	3962	4120	3598	4306
Australia	1873	1660	1630	1620	1682	1621
US	3970	4145	3478	2670	1870	1367
Rest	6421	6127	6105	5879	6283	6618
EU %	42.9	41.0	47.7	53.6	51.4	48.1
	22084	21251	22904	24912	21376	20172
Total	28505	27378	29009	30791	27659	26790

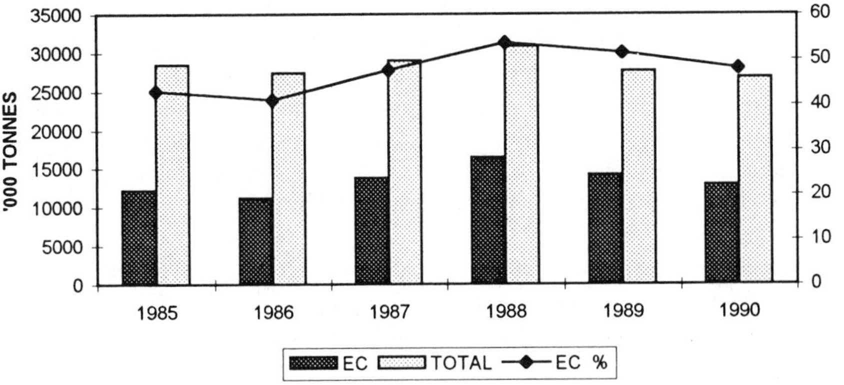

Figure 5.3 EU share of world dairy exports

Note: EU delivery quotas imposed in 1984

other agricultural exporting nations during the 1980s and 1990s. It is the major reason why agriculture was the main issue at stake in the Uruguay Round negotiations initiated by the General Agreement on Tariffs and Trade (GATT) in 1986 and not completed until nearly seven years later in December 1993.

The development of western Europe's role as a major agricultural exporter is almost exclusively a creation of the CAP. Prior to 1972, the nine countries which subsequently made up the EC9 were together net importers of butter and milk powder, net importers of beef and other meats and substantial net importers of feed and food grains.

In the 1980s the EU came to dominate the world dairy market – being responsible for more than a half of all the dairy products in trade – is a major destabilising factor in the international beef market, has the world's largest export surplus of sugar (apart from Cuba) and is a substantial net exporter of grains, with a 21 per-cent share of the world wheat market.[2] The development is succinctly summed up by Professor Alan Buckwell:

> The picture is very clear. In every case, exports have risen. The time path is erratic but the upward trend is unmistakable. Imports have, over the same period [of the development of the CAP in EC9, 1973–85] been static or falling. Thus the balance between the two, the net export balance, has been growing very rapidly. The point at which the Union switched from being a net importer to a net exporter was 1979 for cereals, it was earlier for sugar, in 1975; for wine it was 1976 and for beef and veal 1979. This transition is seen by the rest of the world as an important effect of the Common Agricultural Policy.[3]

Individual EU countries gain substantially from the EU system. Among the EU countries the primary food-exporting countries are the Netherlands and France, who, according to the GATT,[4] steadily increased their share of world food markets by 7.5 and 9.5 per cent respectively in the 1970s and 1980s. Even traditional food-importing countries, Germany and the United Kingdom became fourth and fifth in the world food-exporting league table. The United States, on the other hand, lost market share though remaining the major exporter. The US share of the world food market fell from 18 per cent to 14 per cent, with the value of exports showing only a small increase to $42.8 billion.

These are the sorts of figures which heighten American consciousness of the increasing competition from what the US food and farm industry sees as 'unfair subsidised competition' from Europe. It also provokes continuing resentment among the farmers and food exporters of Australia, New Zealand and Argentina, where farmers are largely unsubsidised. Because of their lack of government protection, the price which farmers in these countries receive is generally the world price: if the EU and the United States depress world prices by subsidising their exports, then their prices and incomes also fall directly - without any compensating subsidies to buffer the impact on their incomes.

The expansion of EU food exporting during the 1970s and 1980s is largely due to one factor alone: the high prices that the CAP has sustained and which have thus encouraged EU farmers to expand output and food traders to export. Farmers and food traders know that Brussels will always pay the subsidy necessary to make the EU's over-priced produce competitive on the international market, so there is no reason not to produce more than the internal market can absorb.

The high prices consistently paid for EU agricultural products has steadily stimulated increased production during the last twenty years. For milk, cereals, beef and to a lesser extent, poultry and pig meat, this has meant that an increasing proportion of the Union's production has become surplus to domestic consumption. The excess has to be exported. There is of course nothing inherently wrong in such a development, but the problem for the Union's trading partners is that because of the high prices paid to EU farmers – normally 35–45 per cent above the world price – the excess can only be

EC share in expansion of farm exports, 1973–82

'000 tonnes	*EC consumption 1982*	*World trade 1982*	*Increase of 1973–4 to 1981–2*	*Growth in world trade 1973–4 to 1981–2*	*EC share change*
Wheat	44400.00	96900.00	12200.00	27900.00	44
Other grains	68600.00	91600.00	9000.00	27300.00	33
Sugar	9400.00	19900.00	3700.00	4400.00	84
Butter	1360.00	780.00	250.00	280.00	89
Cheese	3670.00	800.00	170.00	320.00	52
Beef	6620.00	3860.00	770.00	810.00	94

Source: Bureau of Agricultural Economics (ABARE) Canberra – Agricultural policies in the European Community

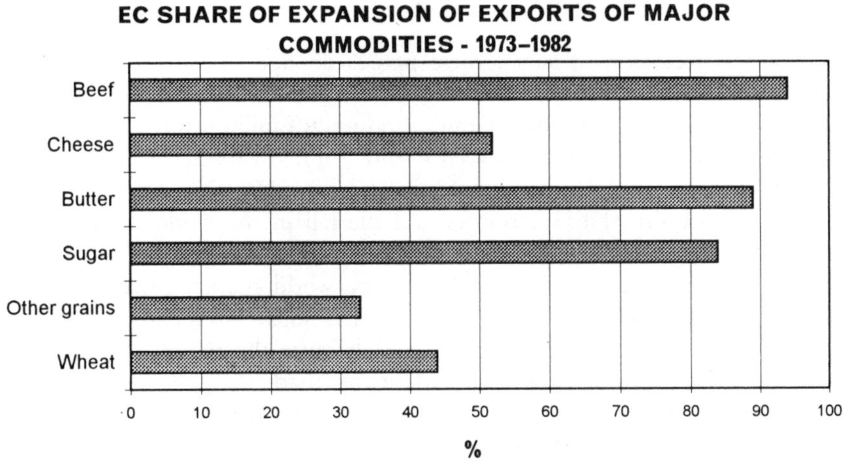

EC SHARE OF EXPANSION OF EXPORTS OF MAJOR COMMODITIES - 1973–1982

Figure 5.4 EC share in expansion of world agricultural trade, 1973–82

exported with the aid of export subsidies. Thus the Union not only stimulates excessive output, but then also subsidises its 'dumping' on the international market.

Over the lifetime of the CAP, the amount of taxpayer's money spent on such export bounties has increased from less than $US500 million in 1970 to a peak of over $US23 billion in 1992. Whereas in the late 1960s the countries which make up the EU10 were net importers of dairy products, grains, meat and sugar, they now export (net) 14 per cent of total milk production, 17 per cent of cereal production, 8 per cent of beef and more than 35 per cent of sugar production. These figures are unlikely to be significantly changed by either the 1992 CAP reforms or the 1994 GATT agreement until well into the next century.

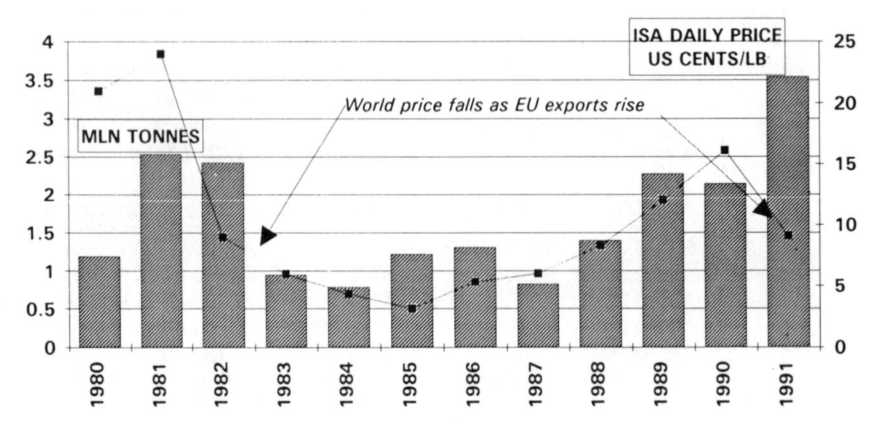

Figure 5.5 Relationship between EU surplus sugar production and exports and the world price of sugar

With the exception of sugar, these figures do not look exceptional, but the actual amounts are enormous and their effect on international markets devastating: 14 million tonnes of milk product, 40 million tonnes of grains, 500,000 tonnes of beef and between 4 and 5 million tonnes of sugar.

With most of the major products, a direct relationship can be traced between the export of EU surpluses and the fall in the world price of the commodity. Sugar is the most dramatic example. As the quantity of surplus Union sugar exported rises and falls, so the world market price falls and rises;[5] this is because the subsidies paid on EU sugar production allow EU exporters to undercut other exporters and because the quantities exported from the Union form such a large part of free market supplies.

The importance of these production–consumption relationships within the Union to international trade arises from their importance in total world trade in the individual commodities: Union dairy exports – 50+ per cent of total world trade, beef – over 15 per cent of beef traded internationally; and sugar, almost a third of this product available on the free world market. In each of these three cases the subsidised EU exports are regarded by international trade experts[6] and other exporters as major, often the major, destabilising force in the market.

In the wheat trade, the EU's share of the market has increased from less than 12 per cent in 1970 to more than 21 per cent; at the same time, the massive expansion of EU production has progressively excluded the wheat and animal feed grains once imported by the EC9 from other grain-producing countries. EU consumers are thus denied access to cheaper food supplies and forced to pay the prices needed to keep the EU's relatively less efficient farmers in business.

Although the European Union frequently denies that one of its farm policy aims is the achievement of self-sufficiency in every food commodity, this is

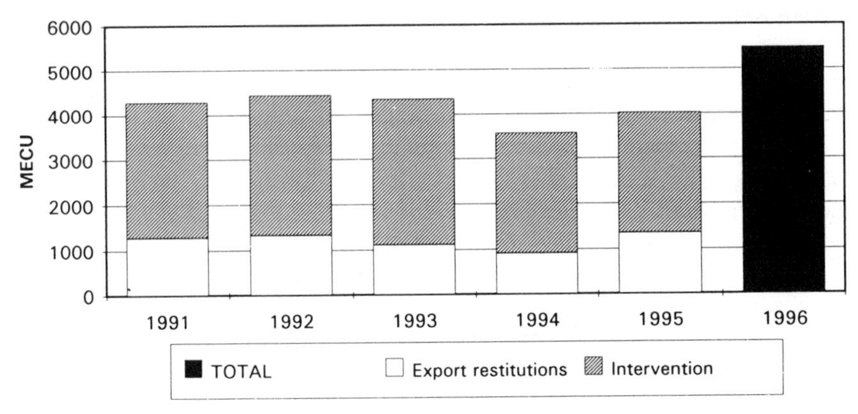

Figure 5.6 EU beef market support cost

certainly how it appears to outsiders. Although the Union is the largest food importer in the world, its imports are mainly of commodities which it cannot produce itself. For all of the main temperate zone foods the Union is self-sufficient or super self-sufficient; where the Union was in deficit in a product, new policies have been introduced to encourage domestic production and crowd out imports.

In 1980, for example, the Union did not have a policy for lamb and mutton. Production, mainly in Britain and France, represented about two-thirds of consumption, with the rest being supplied by third-country producers in New Zealand, Australia and South America. Introduction of the so-called EU 'sheep regime' in 1980–81 led to an immediate increase in prices of 15 per cent: in the following ten years EU production increased by more than 20 per cent. Imports declined by 25 per cent.

But at least lamb is a product in which EU farmers do have a reasonable comparative advantage. This is not the case with some of the other products which the Union's protective policies have encouraged. Take soyabeans, for example: experiments in the 1960s indicated that this crop could be grown in the EU – but at about three times the cost of production in the Midwest of the United States or in South America. This did not, however, stop the Union setting up a market support system. As a result, the European Union now produces close to 2 million tonnes of soyabeans a year. At about the same time it also established market support for rapeseed and sunflower – all crops that cost far more to produce in the EU than in countries with greater technical and climatic advantage.

Paying EU farmers to produce these high cost crops, despite CAP reform, despite the GATT agreement, still costs the Union taxpayer more than $8 billion a year (compared with less than $500 million in the mid-1970s).[7] Not only that, but the oilseed support scheme brought the Union into direct conflict with the United States. As a result of complaints from Washington that an original agreement which allowed free flow of US soyabeans into the EU market without their having to overcome import taxes or unfair subsidies has been breached, the Union was twice condemned by GATT inquiries into the trade-distorting effects of its oilseeds policies.

It is this overspill of its profligate policies into the international market that brings the Union into fiercest conflict with the outside world. Increasingly, the Union has challenged other exporters in their 'traditional' markets with its burgeoning subsidised exports. During the 1980s the EU export subsidy on wheat, for example, has averaged more than $US120 a tonne – that is, often larger than the total world price itself.[8] Generally, the Americans have combated this development with increasing export-subsidy programmes of their own. Less well-heeled countries such as Argentina, lacking industrial wealth with which to subsidise their farmers, cannot, however, afford this luxury.

The first non-EU countries to feel the effect of CAP proliferation were the less developed countries which had traditionally supplied food to Europe, plus developed country exporters like New Zealand, Australia and Argentina. When Britain joined the EC in 1973, for example, it imported 170,000 tonnes and more of butter from New Zealand and Australia and close to 100,000 tonnes of cheese from these two countries plus Canada; it also imported 500,000 tonnes of sugar from Australia. Today, Britain imports only 55,000 tonnes of butter from New Zealand, imports a mere 10,000 tonnes of cheese from the Commonwealth countries and imports no Australian sugar.

Under the CAP, Britain has itself gone from being a large food importer to a significant exporter. As a result of the application of the CAP to British farming, the price of wheat increased from £32 a tonne to over £100 a tonne between 1972 and 1979. As a consequence, some 300,000 hectares of pasture land in Britain was switched out of its traditional livestock farming role, ploughed up and became dedicated to swelling the EC's grain surplus. British banks fell over themselves to lend money on the security of CAP-inflated land prices to fuel the Brussels-financed bonanza with new machinery, chemicals and know-how.

British grain production very nearly doubled, increasing from 12 to 22.5 million tonnes plus between 1972 and 1984.[9]

> Between 1973 and 1978 the volume of United Kingdom agricultural gross output increased by over a fifth. . . . Our self-sufficiency for indigenous foods has risen from 60 per cent to over 80 per cent in 1984. . . . Although largely deriving from technical advances, this progress was powerfully under-written by CAP support policies as they applied in this country.[10]

The 1970s also saw the origins of the agricultural trade war that has smouldered and flared between the EU and the United States over the last ten

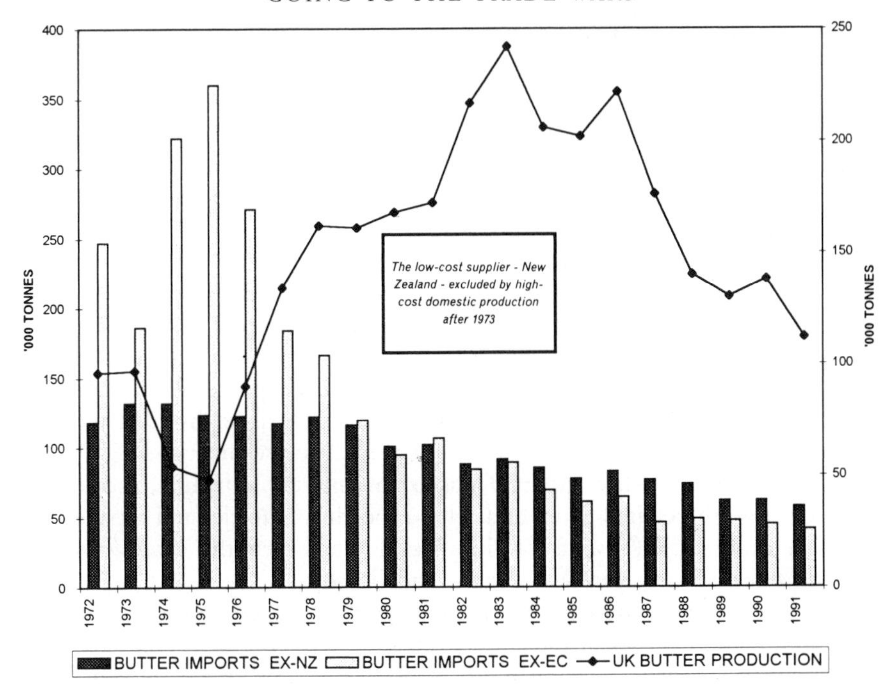

Figure 5.7 Impact of EC membership on UK butter production and imports

years. As the Europeans increased their grain production, so US deliveries of wheat and feed grains were gradually pushed out of the European market. American corn (maize) exports to the EC9, for example, declined from close to 13 million tonnes in 1972 to less than 5 million tonnes by the end of the 1970s. Subsequently, as the EU's production of grain has expanded, the Europeans have come to challenge the Americans in what the United States regards as its rightful dominant place in the world grain market.

This threat of increasingly protectionist European agriculture policies was recognised by powerful American agricultural exporting interests, before its importance was fully recognised by the US Government. Such companies financed think tanks and pressure groups in Europe designed to challenge the increasing drift to what was seen as near-autarky in agriculture. The world's largest grain trading company, Cargill, responsible for approximately 70 per cent of world trade in cereals, even set up its own anti-CAP office in London. The Trade Information Bureau, as it was called, pumped out anti-protectionist propaganda as its wily Director, William Pearce, paced the corridors of power in London, Brussels, Paris, Bonn and Rome, his political antennae constantly on the alert for the latest ramifications of the proliferating CAP.[11] People like Pearce did their best to persuade the British people and politicians against the CAP and urged them to do their utmost to undermine it. Their efforts were, however, largely wasted, since the time when a new member state could make any serious difference to the progress of the CAP leviathan were long past.[12]

Throughout the 1970s and 1980s Brussels used its swelling subsidy 'war-chest' to fight long-running battles for access to profitable outlets for its growing grain surpluses in North Africa, the Middle and Far East and in the Soviet Union. These activities ate steadily into the US share of the world grain market. The Americans had expanded their share of the world grain trade to 56 per cent in 1979, through bringing crop land idled under earlier set-aside policies back into production. When the full impact of EU subsidies hit the market, the United States was thus much more vulnerable, with a much greater share of its output dependent upon exports. This potential instability was realised as the quantity of US grain exports fell by 17.7 million tonnes and the value of its export declined by 20 per cent in the 1979–83 period[13] – the time when the EC9 was expanding its production, excluding imports from the United States and other exporting countries and expanding its own exports. The events of this period established the *casus belli* on agricultural trade between Washington and Brussels.

The Americans retaliated by increasing their special export subsidy programmes – in particular the Export Enhancement Programme, specifically intended by Congress to combat EU subsidies. At the recent peak of the export subsidy battle in 1987, Brussels and Washington spent a total of close to $22 billion on subsidising their grain exports.[14] Throughout the later 1980s and early 1990s the annual expenditure seldom fell much below this figure. A large part of this money undoubtedly did little more than cancel out each other's subsidies.

The main theatre of operations in the 1980s was North Africa. Throughout this period, the United States and the Union slogged it out, subsidy for subsidy in the North African grain market, to the considerable cost of both their own taxpayers and other exporters – mainly the Canadians and the Argentinians.

Historically, the United States has always had a substantial share of the North African market, while the EU's share has only grown as its cereal production and exports have expanded since the mid-1970s. The expansion of EU wheat deliveries – always substantially subsidised – from a mere 125,000 tonnes in 1977 to a peak of 3.4 million tonnes in 1980 – stimulated an escalating subsidy war between Brussels and Washington, which meant that the Union's proportionate and absolute share of the North African market has declined sharply since 1984.

From a mere 2 per cent of the North African market in 1977, the EU share at one point in 1983 rose to 42 per cent and it now averages 10–15 per cent. But the world's greatest agricultural empire struck back. Not content to see this market lost, the US Agriculture Department provided whatever export subsidies were necessary to match the barrage of EU export subsidies

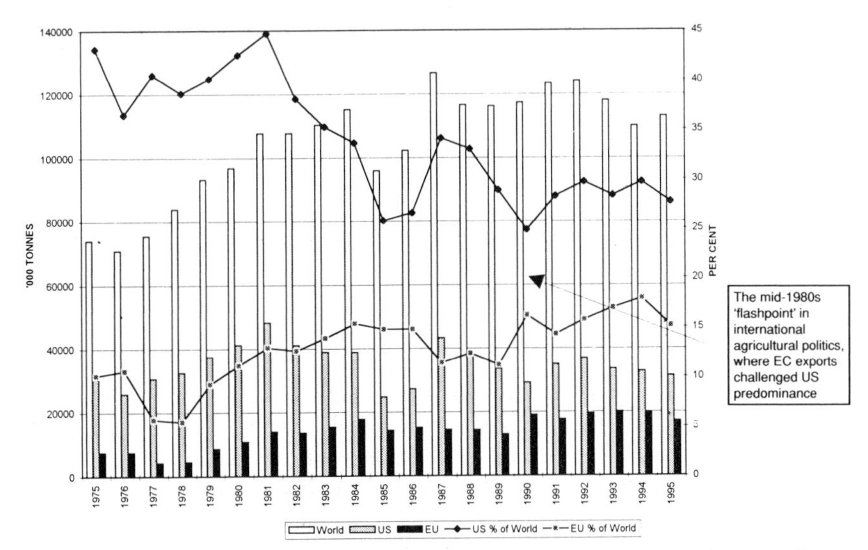

Figure 5.8 The EC challenge to the United States in the world wheat market
Source: USDA and European Commission

and European government-financed marketing aids. The US share of the North African market, while remaining relatively constant to 1983 (45.6 per cent), rose steeply with the assistance of subsidy programmes which targeted the area during the mid- to late 1980s.[15]

Some $1.6 billion was spent by the United States in selling 51 million tonnes of wheat to the Algerians, Moroccans and Tunisians in the 1985–88 period. By 1988, US sales to North Africa represented nearly a third of its total annual wheat export.

But when the giants fight, it's the small operators that get crushed beneath their feet. As the US/EU subsidy war in North Africa hotted up, so Argentina, Canada and, to a lesser extent, Australia got hammered.

Thus for countries outside the EU, the Union's increasingly unacceptable agricultural exporting policy has two main effects: it squeezes them out of the EU and third country markets, and it depresses prices on the international market. Both of these effects reduce the incomes of the agricultural exporting countries.

The EU's role as surplus food producer, exporter and importer is of course highly important to the less wealthy developing countries. As an importer, it provides important concessions to Caribbean exporters of cane sugar and Thai producers of manioc – importing substantial quantities of both these products on preferential terms each year. The cane-sugar producers are, for example, guaranteed the same price as the EU producers for 1.3 million

Table 5.1 North African wheat market shares (%)

	1977	*1980*	*1987*
United States	42	26	61
EU12	2	42	10
Canada	18	10	8
Australia	21	21	20
Argentina	13	0	1
Other	3	0.5	1.5

Source: USDA (from United Nations Trade Statistics)

tonnes of cane sugar delivered to the EU market from the Caribbean and other LDCs in the African, Caribbean and Pacific (ACP) group of countries.

Defenders of the CAP also argue that the EU's role as a grain exporter helps to keep down the price of grain to the poorest nations of Africa and Asia who, at present and for the foreseeable future, are incapable of producing enough to feed themselves. Critics argue, however, that the price-depressing effect of the Union's grain-exporting activities is what holds back many developing countries from increasing their grain output. Governments of these countries import cheap grain, which depresses domestic market prices and discourages their own farmers from increasing their production.[16]

The dumping of increasingly large surpluses on the international market also increases the instability of the international market, thus creating greater problems for countries which have to import from the world market.

It is, of course, not only the EU's activities that have created this instability. The combination of the world-wide application of modern technology and more protectionist government policies is increasing the instability in world grain markets. According to researchers[17] who have examined the phenomenon, while world agricultural production has increased steadily in the post-war period, the nature of that expansion has changed remarkably during the last ten to fifteen years.

Whereas in the 1950s and 1960s world output increased steadily with only small variations around the trend line, in the 1970s, and, even more, in the 1980s, there were increasingly wide variations in output from year to year. This created greater uncertainty and price fluctuation than in the earlier periods.

World wheat prices, for example, remained remarkably stable, at around $60/tonne, up until the early 1970s. The first oil price and general commodity crisis of 1973–75 then set off an instability that appears to have remained and accelerated ever since. Wheat prices tripled by the end of 1973. Since then there have, however, been considerable reductions: the average price of wheat in real terms in the period 1976–87 has been on average only 62 per cent of the average during the 1960s. Prices rose again in the early

1990s under the influence of policy limitations in the United States and the EU combined with huge, subsidised stock-clearance programmes in the 1992–95 period.

The extent of variation in price has risen in the same period: from a variation about the mean of 18.5 and 21.7 per cent in the two decades to 1972 to 35.2 per cent in the 1980s. It is not without significance that the period of greater instability and larger reductions in price have coincided with the period of expansion of EU production, increased exports and declining EU imports.

This close relationship between EU cereal production and world market prices derives from both the level of EU production and the consequent working of the Union's cereal market regime. This is mainly, and obviously, because the Union tends to increase its levels of per-unit export subsidy, its level of exports and its total expenditure on export subsidisation as its production rises – an increasingly vicious circle as rising subsidies chase a falling world price. The effect of this activity is particularly noticeable in the world wheat market. It has been estimated[18] that the elimination of the EU support system would remove a major source of instability from the world market. Although Lingard and Hubbard admit that it is impossible to blame instability on the EU alone, they calculate that if the EU were to dismantle its support system and take to free importing and exporting, then world prices would rise by some 9 per cent. A move to universal free trade has more recently been calculated by the SWOPSIM (Static World Policy Simulation Modelling Framework) (US) model[19] to increase world grain prices by up to 12 per cent.

The United States is of course itself not without blame in damaging the trade of less wealthy farm exporting countries. The United States sees itself as the major contributor to stability in the world's grain markets. Its role as a major producer and major stockholder, it is generally believed – certainly in Washington, makes the United States the underwriter of world food security.

This view is challenged, not only by the EU but also by the less wealthy farm exporting countries. Some analysts argue that the United States' dominant role in the world agricultural exporting industry gives it far more power than the EU to influence market developments. They also tend to argue that this power has been misused. An analysis of the US role in the international grain trade in the 1980s, by the Australian Bureau of Agricultural and Resource Economics (ABARE)[20] suggested that the increased use by the American government of public stockholding to maintain farmers' prices, the increase in the value of the US dollar in the early 1980s and the subsequent adoption of an aggressive export policy to aid de-stocking introduced new instability into the world grain market.

The United States normally holds around 40 per cent of the world's stocks of feed grains and 20 per cent of its stocks of wheat. Since a very large proportion of this stock is either held by government or controlled by government, the trading policies pursued by the US Government are more important in the management of those stocks than is the normal working of the market. In the early 1980s, following the period of sustained expansion of US and world grain production in the preceding decade, there was a steady accumulation of stocks as the Government sought to maintain farmers' prices by buying up surplus grain. This, combined with the relatively high value dollar meant that Washington was effectively setting the world price for grain – a price too high in relation to underlying depressed demand, according to ABARE.

Though US farm price support rates were not raised in US dollar terms, the appreciating US dollar effectively raised them in terms of competing countries' currencies. These countries were thus encouraged to expand production because of the false price signal set by the United States. Thus, according to the Australian argument, the Americans were largely responsible for creating the serious grain surplus problem which led to the subsidy wars of the mid- and late 1980s.

The United States, argued ABARE, should have cut its farmer support levels by a much greater amount in 1984–85 and thus have reduced its accumulation of stocks. Instead, the Government was bidding for farmers' production against the proper market and, in particular, potential export markets and thus forcing up the world price in a period of surplus and therefore encouraging world grain production at too high a level.

Having created the surplus, the Americans then took fright and dumped the accrued surplus on the international market with disastrous results. The US Government subsequently began a policy of de-stocking at prices even lower than they need have been, by using higher than normal export subsidies. This resulted in a complete reversal of the world grain market situation in the 1986 to mid-1988 period: an increase in protection of grain markets in those countries which could afford it and a rapid cut-back in countries more subject to the chances of the world market.

This fire sale also coincided with a change of political philosophy in Washington. The US Government's mighty agriculture policy directorate, the Department of Agriculture, came into the hands of a new Agriculture Secretary – Clayton Yeutter. A convinced free marketeer and proponent of a sophisticated version of Reaganomics, Yeutter was determined to dismantle what he saw as the massive over-subsidising of farmers.

Though among the world's most efficient, US farmers were in the mid-1980s still supported by the US taxpayer to the tune of more than $10,000 each. To dismantle this ornate and expensive farmer support system it was, however, necessary to 'level the playing field' by persuading other developed countries to stop subsidising also, or at least diminish, the subventions to their farm industries. This was the economic philosophical undercurrent carrying along the development of the American position for the Uruguay Round of GATT trade talks which began in 1986.

It was the growing waste and increasing futility of the farm trade war that stimulated the Americans and the other developed country exporters, plus reluctantly – very reluctantly – the EU, to get round the table at the Uruguayan seaside resort of Punta del Este in November 1986 and decide to negotiate a new deal for world agricultural trade that would end this expensive and long-running farce.

Reform of the rules of world agricultural trading – or rather, the imposition of rules on farm trade, because their absence is the root of the problem – was a major objective of what became known as the Uruguay Round of the GATT. Not surprisingly, there was and remains, even after the Round was completed in 1993–94, a vast difference of view between Brussels and Washington on what the hoped-for new world farm trading order should look like.

The eventual American objective was a complete dismantling of all agricultural support, protection and, in particular, subsidisation of farm exports within ten years. Though US negotiators subsequently climbed down from what became known as this 'zero option' (in agricultural trade war terminology, substitute dismantling of export subsidies for dismantling of missile sites), this still remains the longer-term objective of US policy.

The EU, on the other hand, wanted 'market management'. The stonewalling Brussels negotiators argued throughout the long, eight-year (1986–93) Uruguay Round negotiation that free trade in agriculture is a nonsense: it leads only to violent market oscillations, they say, which damage not only farmers but also consumers. Much better, they argue, for the 'big boys' to get together and agree to control the flow of agricultural products onto the market by imposing production quotas, paying farmers not to farm through schemes designed to set farm land aside from production and thus retain the basic means of protecting and supporting farmers' incomes.

According to this EU philosophy, 'zones of influence' should be parcelled out among the major exporters with agreed limitations on subsidies and trade deals. In this way, world farm commodity prices will be kept up and the bill for subsidies minimised.

To the EU this was a much more satisfactory arrangement than what it sees as the 'unmanageable' oscillations of a free market. Officials of the EU Commission's Agricultural Directorate have been among the firmest opponents of *radical* CAP reform. According to their thinking, the basic principles of the CAP – most importantly, the right to subsidise exports – are inviolate and cannot be changed to suit international pressures. The Union must be free to maintain what it calls its 'dual price system': the right to maintain a higher price within the EU than can be obtained on international markets.[21] From this it follows that the Union will only reluctantly give up taxing food imports or subsidising food exports in the immediate future. The terms of the 1994 Uruguay Round conclusion agreement allow it to continue subsidising both production and exports on only a slightly lesser scale than in 1994 until at least 2001 (see Chapter 7).

Equally determined, at the outset of the Uruguay Round negotiation, that the Europeans should change the CAP, was the US Department of Agriculture and the representatives of America's farming and food-exporting industries. According to the Americans, arguments for free trade are irrefutable. Consumers and taxpayers are clearly better off and, in the longer run, farmers too benefit from open markets and unsupported food producers. The problem is, in the short run, that liberalisation also brings pain and misery for those who cannot compete. Although Washington recognised that the EU would not dismantle its CAP – even after any GATT agreement – the main thrust of its Uruguay Round negotiating campaign was that frontier protection and export subsidy expenditure should be cut by 36 per cent over the later 1990s. After four years of largely unproductive wrangling, the Uruguay Round negotiations finally coalesced into a plan put forward by the then GATT Secretary, General Arthur Dunkel.[22] The important features of the Dunkel proposals were: a 36 per-cent cut in budgetary expenditure on export subsidies, a 24 per-cent reduction in the volume of subsidised exports and a 20 per-cent scaling down of domestic supports – all on the basis of the average levels operating in the period 1986 to 1990.

The Dunkel Plan recommended that import levies, export subsidies and other, non-tariff barriers to free trade should be reduced by converting all barriers to a 'tariff equivalent' which would then form the basis for reductions. Dunkel recommended that this process must result in a minimum reduction in the tariff of any individual product of 15 per cent. Similarly, the plan recommended that there must be a minimum increase in import access of any products of 3 per cent, rising to 5 per cent by the end of the initial GATT agreement implementation period. With considerable modification, the Dunkel Plan became the basis for the agreement that was eventually reached at Geneva in December 1993.

The important feature of the Dunkel Plan, though rejected by the negotiators, was that it effectively provided the means of bridging all but one of the main gaps between the EU and the United States – the two major protagonists in the long-running farm trade argument. Throughout the endless and often tedious bilateral negotiations between the two – which had overshadowed the Uruguay Round since its very beginning – the question of subsidised exports predominated. Despite the fact that 108 countries were involved in this negotiation, it is generally acknowledged that had this dispute between Brussels and Washington not been settled, there would have been no Uruguay Round agreement. Even the group of fourteen major agricultural exporting nations, the Cairns Group[23] led by Australia, were forced to sit on the sidelines while the two champion subsidisers slugged it out.

The United States was determined to impose limits on the EU's burgeoning wheat exports, while the EU, though acknowledging that it would have to give some concession in this area, was equally determined to reserve the right to export as large a quota of wheat, other grains and grain products as possible. It was also determined to maintain its dominant position in the world dairy market. Though initially not prepared to concede any undertaking to limit the quantity of exports but determined to concede only on the level of budgetary expenditure, the Union had to make concessions by reducing the money which it spends dumping its exports on the international market.

The final argument refined itself into a US demand for a reduction in EU wheat exports to 11 million tonnes, while the EU refused to drop below 15 million tonnes. Unofficially, it was known that the United States would accept a limit of 12 million tonnes and the EU would concede a limit of 13 million tonnes. Dunkel effectively closed this gap. The formula in the GATT compromise limited EU exports to 12.7 million tonnes by 2000.

As the UR negotiations reached their final stage, it became obvious that the European Union would have to make significant concessions; it was recognised that this could not be done without considerable modification of the EU's domestic farm policy.

The EU Commission therefore recommended in its 1991 proposals[24] for reform of the CAP that cereal support prices should be reduced by a third, while farmers should be compensated through direct subsidies to make up for the lost income. This would have the effect of maintaining the level of support to farmers at the pre-reform level while rendering the CAP – after considerable argument – 'GATT-legal'. Though this stratagem has the effect of largely eliminating the EU's direct export subsidies, it still gives European farmers an equivalent advantage which would allow them to compete unfairly on international markets (see Chapters 6 and 10). This move was rightly opposed by the 'free-trading' agricultural exporters in the Cairns Group and also by the GATT itself. The GATT experts see direct subsidies to farmers, paid on the basis of their production, as a direct distortion of trade and outlawed them in their Uruguay Round proposals. The EU's compensation subsidies were for the purposes of achieving a GATT deal, fictionally regarded as not being 'production-positive'.

Although the 1993 agreement goes nowhere near the radical change which the free traders wanted, it is expected at least to stabilise the levels of subsidisation of farmers. It will also bring about, gradually, a reduction of import protection.

Had there been no GATT agreement on liberalisation of world food policy, then the cost of protecting and supporting farmers and food markets would have continued to increase. No one can calculate precisely what the benefits of a new trade order will be or the losses of continuing with the pre-1993 policies would have been. Only estimates can be made with the aid of computer programs designed to incorporate the many complex factors that have to be taken into account.

According to one of the longest-running and most respected of these studies,[25] the annual cost of agricultural support to the industrial countries would without reform have more than trebled in real terms by the year 2000, compared with the cost in the late 1980s. There would have been an increase in the cost of support of the seven major food commodities (wheat, coarse grains, rice, ruminant meat, non-ruminant meat, dairy products and sugar) to US$57 million by 1995 compared with around $30 billion in 1990. This figure does not include the considerable consumer cost of continuing the pre-1994 policies.

The study suggests that the total benefits to consumers and taxpayers of full liberalisation of world food trade – something which the 1993–94 agreement by no means achieves, would amount to more than $170 billion by 2000 (a gain of $1400 to $2800 for each non-farm household). Farmers in less developed countries would gain an estimated $35 billion, while small farmers in developed countries would bear $20 million of the reduction in payment.

The increase in agricultural support and protection in the 1980s has meant that, whereas producer prices in developed countries were 40 per cent above the unsupported level in 1981–92, they were likely to have been 80 per cent above by 1990. By including such factors as population change, income and real exchange rates, the study projects likely price levels by 2000 on the basis of continued unreformed agricultural policies, or those under reform.

The calculations suggest that while liberalisation of the EU and Japanese agricultural policies would raise world prices most, an all-round liberalisation would be likely to raise world food prices by as much as 25 per cent and by substantially large amounts by 2000. The increases are likely to be most notable for dairy produce and beef, but also for food grains and sugar.

The assessment indicates that the unsubsidised export price of wheat from the EU would have risen (though the internal EU price would be less, the removal of the export subsidy means that the delivered price to the external world market rises) by 15 per cent by 2000, the export price of beef and lamb by 33 per cent and of dairy products by 72 per cent.

> The European Community continues to be by far the most important contributor to these price effects. Even though rates of protection are higher in other west European countries and Japan, those countries are smaller participants in world food markets than the Community and hence their policies have less influence on international prices . . . according to these results, United States policy is still having a much smaller net effect on world food markets than is the policy of the European Community.
>
> (Anderson and Tyers, *Global Effects of Liberalising Trade*)

Had no steps been taken towards liberalisation in 1993, then prices would have risen substantially. Had there been full liberalisation of agricultural trade, the EU and Japan could have been expected to import 37 million tonnes more grain, in the EU's case switching from net exporter to net importer.

US and Australian exports of grains would have increased by 17 and 5 million tonnes respectively.

What is not in doubt is the substantial benefit which would accrue to American agriculture and the US economy as a result of a new world farm trade order. The 1993 agreement on agricultural trade is expected to result in significant increases in US agricultural exports and farm incomes.[26] An analysis by the US Department of Agriculture indicates that an agreement to cut export subsidies by around a third over the five years from 1995 would result in an increase in US export agricultural revenues of $4–5 billion and an increase in farm incomes of $800 to 1000 million by 1988. The 1994 Uruguay Round agreement includes broad assent to a 36 per-cent reduction in export subsidy expenditure in the 1995–2001 period; its net effect is likely to be a reduction of approximately 30 per cent. The USDA emphasises that its estimates are conservative, since they do not take fully into account the effects of world trade liberalisation on demand for food.

The original USDA analysis assumed that export subsidy expenditure would be cut by 36 per cent, export volumes by 24 per cent and domestic supports by 20 per cent, with all tariffied[27] trade barriers being reduced by 36 per cent over the 1993–99 period. Internal production subsidies are assumed to be reduced by 20 per cent.

American farm income gains will come from an increase in market returns which will exceed the losses from the removal of government supports, with receipts increasing by an estimated $4.6 to 5.2 billion and government payments declining by $2.6 to 2.7 billion. After allowing for cost increases of $1.2 to 1.3 billion, net farm income will increase by between $0.8 and 1.2 billion. The USDA says that conforming with the GATT recommendations will involve no significant changes in current US agriculture policy – the required 20 per-cent cut in domestic farm support can be achieved by the changes already being applied under the 1985 and 1990 Farm Bills. The advantage to the US Treasury is further consolidated by the 1995 Farm Bill.

The most significant gains to the US agricultural export industry will come from the increased returns to the cereals sector. Most significant would be the benefit to wheat producers and exporters.

> The United States would benefit from expanded export opportunities for wheat that would lead to larger exports and higher prices. The major factor affecting the world wheat market through 1988 would be a reduction in subsidised wheat exports.[28]

It is not expected that the reduction of US export subsidies would reduce American farm exports, since more would be gained from other countries reducing their exports: by 2000 US wheat exports could be expected to rise 10–12 per cent above the baseline level.

> This boost in exports would raise farm prices by 17–19 per cent and production would increase 1–3 per cent. Wheat producers' gross income

> would increase 3–4 per cent, with market receipts rising and deficiency payments falling because of higher prices. Because the US wheat programme was adjusted between 1986 and 1990 and again by the 1990 Farm Bill and budget legislation, no additional programme changes would be required to meet the cut in internal support.[29]

In the livestock sector the USDA believes that world dairy prices will increase by 10–15 per cent by 2000 – 'compared with prices if current global trends were to continue'. Tariffs would remain high on access to the US dairy market, with imports continuing only under access provisions as at present. These provisions would have to be increased to conform with the GATT provisions for a 3 to 5 per-cent increase in import access. These would be likely to amount to the equivalent of an additional 2.6 billion pounds weight in milk equivalent. These imports would be the key factor affecting US dairy product prices by 1998: producer revenues could be expected to fall by as much as 3 per cent.

The USDA expects a GATT agreement to result in an increased global meat trade arising from two main sources: increased market access to developed country markets and increased demand from less developed countries. Reduced subsidised EU exports of beef and pork would also stimulate demand for US exports. US exports would expand and the US import market would contract. It is expected that US meat exports would increase 10–12 per cent; there would be particular increases in deliveries to Japan.

It does not necessarily follow, however, that the EU agriculture industry should lose as much as the United States appears to be gaining or the European farm lobbies fear. The efficient section of the EU industry which produces more than 80 per cent of the Union's food output is very nearly as efficient as its transatlantic counterpart.[30]

From an environmnetal standpoint, it can of course be argued that EC production is much more 'costly': while the Union's most efficient farmers can probably produce at a lower cost per tonne, North American producers have a much lower cost per hectare. This is because of the very much higher European yields – achieved through heavy applications of fertiliser and crop protection chemicals.

The 1994 agreement is far from perfect, most obviously in the way in which it allows the Europeans and the Americans to continue paying production-linked subsidies to their farmers which have much the same effect in international competitive terms as direct export subsidies. But at least the partial solution embodied in the 1993–94 agreement will mean a start on the road to liberalisation. Much worse would have been a collapse of the GATT talks[31] without even the beginnings of the process of liberalisation which is now taking place.

6

THE GREAT REFORM?

Fitting European agriculture into the world system

CAP REFORM AND THE GATT

What is not in doubt is that European agriculture in the 1990s is at a turning point, both technically and politically. Scientists have now developed and are continuing to develop new techniques, new plant and animal genomes, indeed, new farming systems which will increasingly allow more food to be produced with fewer resources. The main effect of these changes, in the absence of political interference, is to encourage farmers to increase or at least maintain their profits by increasing production. This has posed and is continuing to pose a fundamental question about European agricultural policy: should the state continue to interfere in agriculture and food production and markets or should it leave an untrammelled market to dispose resources and to regulate output?

Broadly, the debate in Europe – as everywhere in the developed world – is between those who say that, left alone, farmers and traders will adjust production to meet the quantity and quality of food demanded and those, on the other hand who say that agriculture cannot be left to the invisible hand of the market, since the fluctuations in production will bring about wide variations in supplies and prices, with both farmers and consumers being consequently harmed. There is a third faction, which may or may not be related to the second of these two broad schools: those who believe that, because agriculture controls such a large proportion of the land area and therefore of the environment, it cannot be left to self-regulation by farmers working through the market.

In the 1990s, the market manipulators, the social engineers and environmentalists will increasingly coalesce into one main group. Those who believe that it is socially necessary to maintain and sustain village communities and the population of rural areas are allying with the traditional market managers who say that agriculture is too important to be left to the rigours of the open market.

On the other side are the agricultural free marketeers. They turn the traditional arguments for intervention on their heads: modern science and agricultural technology means that farmers can now so well regulate their production – even to the extent of being able to override all but the most extreme climatic variations – that it is no longer necessary to iron out the market impact of production variations through official intervention. In other words, say this group, agriculture can now be treated like any other industry. The farmer, thanks to modern science and, in particular, recent developments in biotechnology, is able to adjust his output to the needs of the market just like any other entrepreneur.[1]

Because farms are now much larger than they used to be and getting larger, the farmer is as capable of judging the market and producing the right commodities in the right quantity as the manufacturer of motor cars, washing machines or computers. Given his modern industrial approach to production, what justification, ask the free marketeers, is there to continue interfering in farm markets?

In the European Union these two broad schools of thought are at the root of the seemingly endless arguments about the future of agriculture policy. Although in every member country there are ardent supporters of both – or all three – of the main different views on the future of agriculture policy, they do tend to divide on national lines. In general those countries with a long tradition of dirigiste interference in agricultural markets see it as necessary to go on interfering for social and environmental reasons as well as reasons of food security. The countries where this approach is most deeply entrenched are France and Germany. At the other extreme lies general opinion in the United Kingdom, which maintains that the job of the farmer is to produce food – in as large a quantity and as cheaply as possible and with the least interference from government.

Between these two poles lie countries like Denmark and the Netherlands which are leaders in the environmental regulation of land use, but who believe less in the social role of agriculture and would opt for agricultural free trade if they believed that their efficient farmers would be guaranteed a level economic playing field in the EU and world markets. In France and Germany a well-populated countryside with thriving villages has been sustained by large-scale national and local government intervention and – in Germany's case – by clever manipulation of EU agriculture policy so as to channel support to the smaller farmer.

THE 1992 CAP REFORMS

The increasing cost of the CAP, its tendency to produce large surpluses of major commodities and its failure to support the majority of EU landholders finally forced the European Commission and the Council of Ministers to begin fundamental reform. The emphasis on market intervention was identified as the major cause of inefficiency: supporting markets rather than farmers resulted in support being concentrated upon the 20 per cent of farmers responsible for 80 per cent of output rather than the majority of farmers whose farms are too small to benefit significantly from the market. After several ineffectual reform attempts in the early 1980s, the EU Commission proposed[2] in 1991 and in 1992 the Council approved,[3] a radical change in the method of supporting the Union's 8 million farmers. The reform redirected the emphasis from markets to direct subsidisation of landholders. The overall aim of the reform was twofold: to cut over-production and to maintain rural prosperity by supplementing the incomes of small farmers by direct subsidies.

The EU's Council of Ministers in 1992 agreed this proposed change and thus laid the groundwork for the eventual more profound reform of the CAP. The agreement provides the mechanisms to begin the process of switching the emphasis in the agricultural support system from agricultural markets to landholders.

The Council at last accepted that the open-ended and open-handed support of commodity markets does not give adequate support to the farming population, but does create policy management problems that cannot be solved by mere adjustment of price levels or market intervention conditions. It is, however, unlikely that the uncoordinated combination of price adjustment, direct subsidisation and set-aside agreed in 1992 will reduce cereal surpluses; it is, though, likely to stabilise output. It certainly will not reduce the agricultural support bill. A substantial level of export subsidisation will still be required – unless the new GATT rules on agricultural trade applied from 1995 and possible changes in world population and food demand result in a permanent increase in world prices. Such a development would close the wide gap between the EU and the world price still remaining after the 1992 reforms have been fully implemented in 1996.

The measures for the livestock sector largely continued the traditional system unchanged, with only modest adjustment of the present levels of price support, commitment to market intervention and to the subsidisation of exports. The most important element was the introduction of direct payments to beef and sheep producers based on their flock and herd numbers in the 1990–92 period, combined with reductions in market price support. It is claimed by farmers and their representatives that these limited 'headage' payments represent 'quotas' on beef and sheep production. This is not so. For both of these commodities the reforms introduced limitations on the amount of EU subsidy (based on numbers held in 1990–92) which each farmer can claim on his beef herd or ewe flock: there is no penalty for keeping more animals than are covered by the subsidy, or even for producing without any subsidy at all. A true quota such as the EU milk delivery quota, or national quotas for potatoes and hops, involve heavy penalties – greater than the market price of the commodity – on any production outside the quota.

	000 ha 1985	t/ha Yield	Production	000 tonnes Consumption	Feed	Export surplus
1975	15192	3.15	47882	48717	12074	– 895
1976	16187	3.04	49287	49919	12578	– 632
1977	14683	3.23	47407	51426	12718	– 4019
1978	15749	3.71	58464	51667	13915	6797
1979	15519	3.63	56288	52716	14961	3572
1980	16314	3.96	64639	52944	15146	11695
1981	16326	3.74	61046	53312	15957	7734
1982	16615	4.06	67399	53640	17340	13759
1983	16812	4.01	67339	60065	23346	7274
1984	16964	5.13	87037	64145	25687	22892
1985	16029	4.71	75563	63429	26300	12134
1986	16473	4.63	76228	61654	24432	14574
1987	11630	4.54	75486	63004	24763	12482
1988	16263	4.82	78376	64099	24808	14277
1989	16968	4.84	82047	61825	23890	20222
1990	16517	5.13	84709	62622	25793	22087
1991	16875	5.36	90422	64582	24559	25840
1992	16833	5.04	84776	62292	23054	22484
1993	15111	5.28	79805	69642	29916	10163
1994	15193	5.39	81881	69722	30039	12159
1995	15559	5.46	84965	69390	30265	15575
1996	16492.54	5.54	91400	70084	30568	21316
1997	16492.54	5.63	92771	70785	30873	21986
1998	16492.54	5.71	94163	71493	31182	22670
1999	16492.54	5.80	95575	72208	31494	23367
2000	16492.54	5.88	97009	72930	31809	24079

Figure 6.1 EU wheat production and export projection to 2000

The 1992 reforms, amazingly, made no change at all to the support of the Union's major policy problem area – the dairy sector.

The main emphasis of the 1991–92 reform package was on the cereal sector, with the aim of substantially reducing the level at which market prices would be supported and thereby, it was intended, diminishing the Union's commitment to subsidise exports, while at the same time seeking to stabilise production through setting aside a certain percentage of the Union's arable land from production.

The new arrangements can be summarised as a change from a wholly market-supported system to one where the market support is diminished and direct subsidies and set-aside play a much more important part in the support of the producer and the regulation of the market. In essence, under the new system the target market price level is being cut by almost a third over the period to 1997. Individual producers are being fully compensated for this potential loss of income through direct compensatory payments equal to

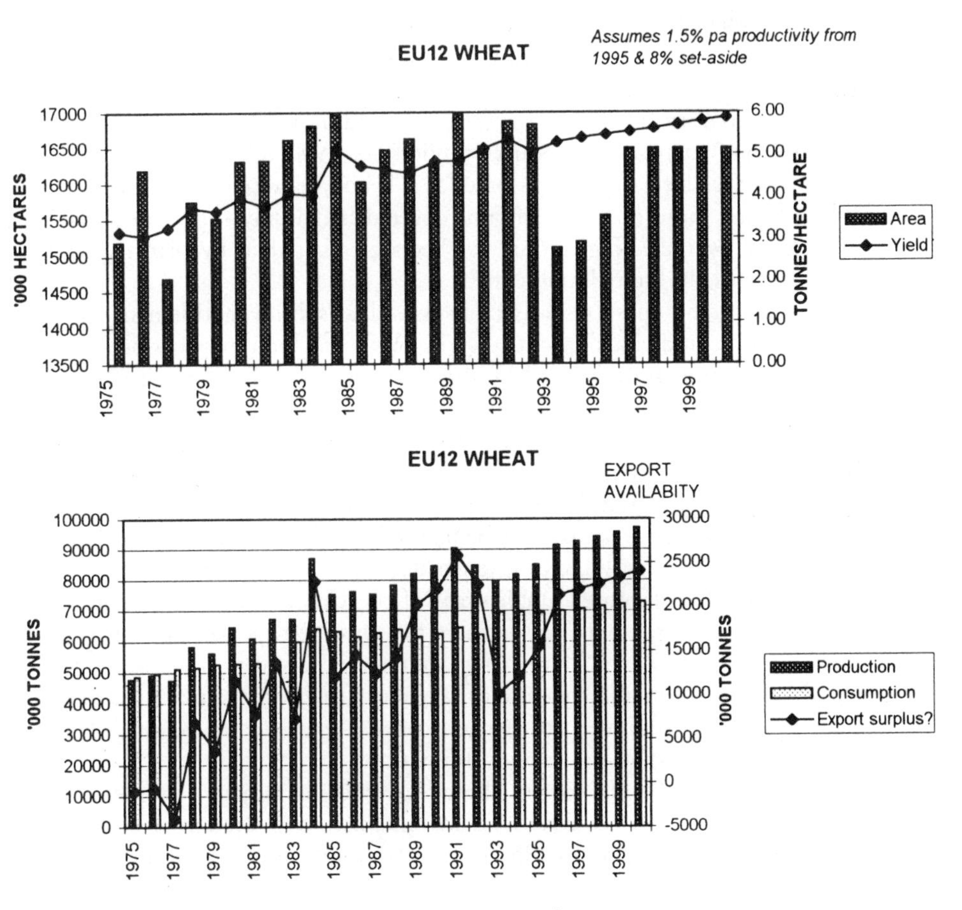

Figure 6.1 (continued)

the difference between the current official price and the eventual substantially reduced price – provided that they agree to set aside from production the agreed percentage of their arable land. In the first year of the new system, 1993–94, the amount which had to be idled from production was 15 per cent of the farmers' average arable area.[4] Compensation for the reduction in market support is, however, only paid on the basis of pre-1992 average regional yields.[5] As a penalty for over-planting, set-aside is increased to the extent that the total regional production exceeds the starting area (the 'regional base area') multiplied by the average regional yield: each 1 per-cent excess results in a 1 per-cent increase in set-aside in the following cropping year.

Under this scheme individual producers will only receive compensation up to the level of their officially registered arable base area (effectively their 1992 area) multiplied by the average regional yield multiplied by the per tonne compensatory payment. This means, therefore, that the amount of production on which the individual producer will receive the full return – market price plus compensatory subsidy – is effectively fixed at 1992 levels. As with the new livestock payments introduced under the 1992 reform, farmers argue that this is effectively a quota. In practice it is not, because there is no penalty for producing more than the amount covered by subsidy – provided that it is produced by increasing output per hectare rather than by planting more than the officially limited arable area.

Two important questions arise: (1) will this 'subsidy entitlement quota' mean that the individual producer will in future limit production to the amount of output fully covered by the compensatory subsidy; and (2) what effect will set-aside have in reducing production?

The answer to the first question is that, for the 20 per cent of grain growers who produce 80 per cent of the Union's output, it is still profitable to produce beyond the subsidy quota – provided that the extra production comes from increasing yields (the area that may be planted is of course officially restricted). With future productivity increases estimated for wheat at beyond 2 per cent a year in the most favoured arable areas of the Union and with current cost levels in those areas, it will pay farmers to continue to expand production.[6] This initial conclusion by farm management economists has been subsequently supported by the response of efficient farmers in the main EU arable areas: they have continued to increase cereal – particularly wheat – yields and production despite the subsidy limitations.

Productivity is also the key to the effectiveness or otherwise of set-aside. The decision to allow set-aside land to remain in the rotation will in itself ensure that production on the 85 per cent (in 1995–96, 92 per cent) of land which is not set aside, but remains in production, becomes more productive. American experience[7] indicates that the concentration of the individual farmer's management and physical resources on a smaller area of land immediately results in an increase in production on the farmed area; the net output thus falls by considerably less than the percentage reduction in cultivated area. In addition, an increase in productivity of 2 per cent (for wheat) would eliminate the effect of a 15 per-cent set-aside in four years; combine both these factors and it is unlikely that the reduction in output resulting from a 15 per-cent set-aside will be more than 5–8 per cent. In practice, the 15 per-cent set-aside of 1993–94 resulted in a reduction in total cereal output of 8–9 per cent; production of the most important commodity in surplus, wheat, fell by only 4 per cent.[8]

The flaw in the post-1992 arrangements is that they sustain profit levels of the larger and more efficient grain producers at too high a level, while not providing adequate curbs on increased production. The effect is likely to be that cereal production will tend to be maintained at early 1990 levels and will probably increase in the later 1990s.

Box 6.1 Why EU cereal production will continue to increase

The European Union's production of cereals will continue to increase, despite the 1992 reform programme, for three reasons: (1) the growing majority of efficient farmers will still find it profitable to increase production; (2) the apparently most radical check on production incorporated in the reform programme – land set aside – is a limitation on planted area, not on total production; (3) productivity. Although the reforms have cut the official support price by a third (over the 1993–96 period), compensatory subsidies make up the difference on the individual cereal grower's pre-reform (1992) level of yield. The combined total of lower officially supported market price and the subsidy still gives the farmer the same return per tonne as before the reform. The important difference is, however, that because the subsidy is paid only on the farmer's historical yield (officially, his 'reference yield'), any increased production receives only the much reduced market price – without any make-up subsidy.

The official view is that this reduced price is below the cost of production of cereals and that therefore, because no subsidy is paid on any increased output, farmers will not produce more than the amount upon which subsidy is paid. And because the area that may be planted is restricted by set-aside, production will not increase. There is, however, a fundamental flaw in the official reasoning: that production of the 'marginal' tonnes beyond those on which subsidy is payable is unprofitable. This is not so. While it would possibly not pay the majority of EU cereal growers to produce grain at all at the new low post-reform price if no compensating subsidy were paid on any part of a farmer's output, the total return received under the 1992 reform programme (lower price plus subsidy) on the bulk of production (the reference yield) covers all the farmer's fixed costs and most of his variable costs of production. The extra tonnes above the 'subsidy limit' require no more machinery, no more labour and no more land. All they do require is, possibly, more fertiliser and more crop protection chemical. Even this is doubtful, since most of the increased output will come from more skilful husbandry and genetically improved wheat and other cereal varieties – in other words, from increased productivity. Such productivity increases can generally be achieved without increased chemical use. What is important, however, is that the extra cost of production of the extra tonnes of output above the subsidised level will always be less than the return from the market. It will therefore pay efficient farmers to grow more cereals – even though they will not be able to draw subsidy on the additional output. Production will continue to increase.

It is therefore likely that, by the late 1990s, the EU will still have wheat production close to 90 million tonnes, will still have a large exportable surplus and a commitment to subsidise the export of at least half of its surplus.[9]

Almost the whole of the Commission's reform proposals for the dairy sector were either rejected or diluted by the Council of Ministers. The Council failed to cut quotas and cut prices by minimal amounts, and thus left the sector largely untroubled by change. Totally ignoring the implications of the impending GATT agreement, the Council decided that the relative buoyancy of the world market in 1992–93 indicated that they did not need to cut either the 100-million-tonne-plus milk delivery quota or to cut intervention prices for the major surplus product, butter – despite the fact that milk production is currently 20 per cent above domestic consumption, the disposal of close to 550,000 tonnes of butter has to be subsidised each year and that the dairy sector costs the taxpayer an average 5 billion ECU a year.

The Commission had originally proposed that the milk quota should be cut by 4 per cent over the three marketing years 1993–96, with 1 per cent of the cut being put into reserve for redistribution among producers. This would have given a net cut of 3 per cent by 1996. This reduction of 3 million tonnes of milk would have cut 105,000 tonnes off the butter surplus and 234,000 tonnes off the annual million tonnes of skim milk powder which has to be subsidised onto the animal feed and export markets. In addition, the Commission had proposed that the intervention price for butter should be cut by 15 per cent over the three years and skim milk powder by 5 per cent, to give an overall milk target price cut of 10 per cent by the end of the third marketing year of the reform plan in 1995–96.

The Council agreed four main reform measures for the beef sector: (1) a 15 per-cent cut in the official buying-in or intervention price for beef carcasses; (2) a physical limit on the quantity of beef which the EU authorities are allowed to buy into intervention storage – 750,000 tonnes in 1993, falling to 350,000 tonnes in 1997; (3) a 15 per-cent cut in the official buying-in or intervention price over the 1993–96 period, and (4) the introduction of larger direct subsidies to beef producers for each breeding ('suckler') cow and finished beef animal ('beef special premiums' – BSP) which they keep (so-called suckler cow premiums already existed prior to reform, but these were increased as part of the 1992 programme). Payment of the beef special premiums are limited to ninety animals on any one farm (hence the claim by farmers that the new system is *de facto* a 'quota' on beef production).

The main features of the changes in the beef sector are the cuts in intervention price, the limits on intervention purchases and the increased importance of the compensatory beef and suckler cow premiums. The 15 per-cent reduction in the intervention price over the 1993–96 period and the reductions in intervention buying quantities are the major influence on beef prices in the period to 1996, while the significant limitations on subsidised exports in the second half of the decade, arising from the GATT agreement, will have a further substantial weakening effect on prices.[10]

The major effect of the CAP reforms on the beef sector will be to encourage the production of beef on specialist beef farms using suckler cows on extensive grassland, while giving far less encouragement to intensive grain-rearing of beef animals. Suckler cow systems – involving more extensive, grassland-based production methods – will gain most benefit from both the suckler cow and beef special premiums. Large, cereal-feed-based units will be seriously handicapped by the 90 head a year limit on the entitlement to BSP.

The most important development with which the EU beef industry will have to deal is the severe limitation on the subsidised export of beef which will begin to affect the market from the mid-1990s. This is likely to mean a reduction of close to half in the 1992 level of export – 350,000 to 400,000 tonnes by the end of the GATT period (1996–2000). It is difficult to see any way in which the Union could avoid this export reduction.

Despite obvious shortcomings, it cannot be denied that this undoubtedly historic decision by the EU's Council of Ministers laid the groundwork for the eventual reform of the European Union's common agricultural policy. There is, however, still a long way to go before the CAP will no longer distort agricultural markets, eliminate subsidisation and cease being a major burden on the taxpayer and consumer. If, for example, the EU authorities maintain the compensatory subsidies – formulated as a temporary mechanism designed to adjust EU agriculture to producing at or close to the world market price – at the full post-reform levels beyond 1997, not only will production continue to increase, but so too could EU exports or stockpiles.[11] A large CAP budget burden will persist.

This is because the cost of the compensatory, direct-subsidy payments which have now replaced a large part of the export subsidy and intervention payments will cost more than the old system.[12] What the 1992 CAP reform programme does is to provide the mechanisms to begin effective reform: the EU authorities have to go a great deal further before consumers, taxpayers and third-country food exporters will see any major benefit.

Most importantly, the CAP reform has to be reconciled with the 1993–94 agreement in the GATT.

Box 6.2 The 1992 CAP reforms in the cereals sector – detail

Because of their central role as human and animal food, the emphasis of the 1992 reform programme is on cereals and arable crops. The assumption is that reduction in grain prices will force an overall moderation of EU farm prices, since grain forms the raw material for the production of most livestock products and sets the relative profitability for the arable farming sector. Reduction of cereal support levels to a point closer to the world price will, it is expected, provide an overall movement of food prices towards a more realistic market-clearing level. Cereal support prices are being cut from the 1992–93 effective support level of 155 ECU/tonne to 110 ECU by the 1996–97 marketing year.

Farmers with less than 20 hectares (producing less than 92 tonnes/hectare) and those who sign on to set aside the officially designated annual proportion of their arable area (15 per cent in 1993–94, 12 per cent in 1994–95 and 10 per cent in 1995–96) will be paid compensatory subsidies equal to the difference between the 1991–92 support price and the new lower support price – calculated on the basis of the average regional yield in the years to 1988–92 (excluding the lowest and the highest). Thus any increase in yield on this level will not receive compensation.

Table 6.1 New EU institutional and producer price levels ECU/tonne for wheat (compared with pre-reform – 1992–93)

	1992–93	*1993–94*	*1994–95*	*1995–96*	*% Reduction 1995–96/ 1992–93*
Notional target price[1]	–	130	120	110	–
Effective intervention	155	177	108	100	35.5
Expected market price	155	130	120	110	29
Compensatory payment	0	25	35	45	–
Producer return[2]	155	155	155	155	–
Threshold price	221.68	175	165	155	30.1
Union preference[3]	c.62	45	45	45	27.4

The *institutional* prices (target, intervention and threshold) are set by the EU Commission and Council each year at a level designed to support farmers' incomes at an adequate level. Exceptionally, in 1992 prices were agreed for three years ahead through the reform period. This table shows the change in emphasis from market support (the effective intervention price) to direct subsidisation of producers (the compensatory payment).

1 The target price in the new system is not the same as in the old: the old TP was an unrealistic figure set above the threshold; the new TP is a realistic assessment of the level at which the internal EU market price should settle.

2 Note that while the producer return per hectare does not decline under reform, because the reduction in market price 1993-6 is matched by the compensatory payment, total return per producer falls because 15 per cent of area is taken out of production by set-aside (this will vary according to the annual set-aside requirement).

3 Under the old system, Union preference was the difference between the *official* intervention price (159.50 ECU/tonne) and the threshold price; under the new, it is deemed to be the difference between the target price and the threshold.

Prices: ECU/tonne

	1991–92	1992–93	1993–94	1994–95	1995–96
Threshold	220	210	175	165	155
New 'target'			130	120	110
Intervention	155	150	115.5	106.6	98.71

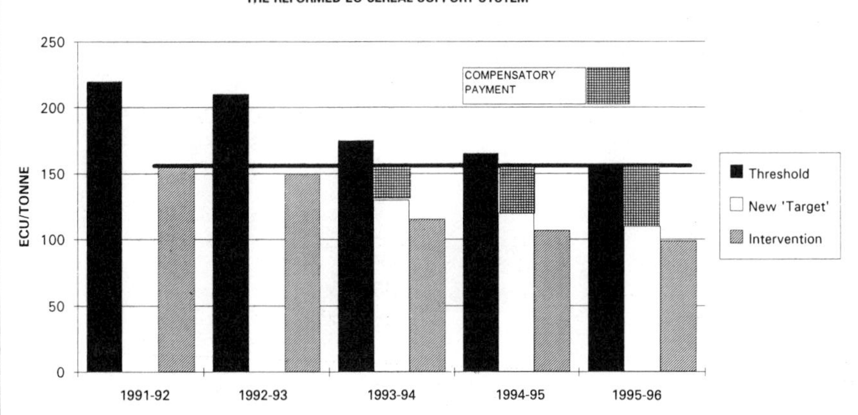

Figure 6.2 The new EU cereal market support system

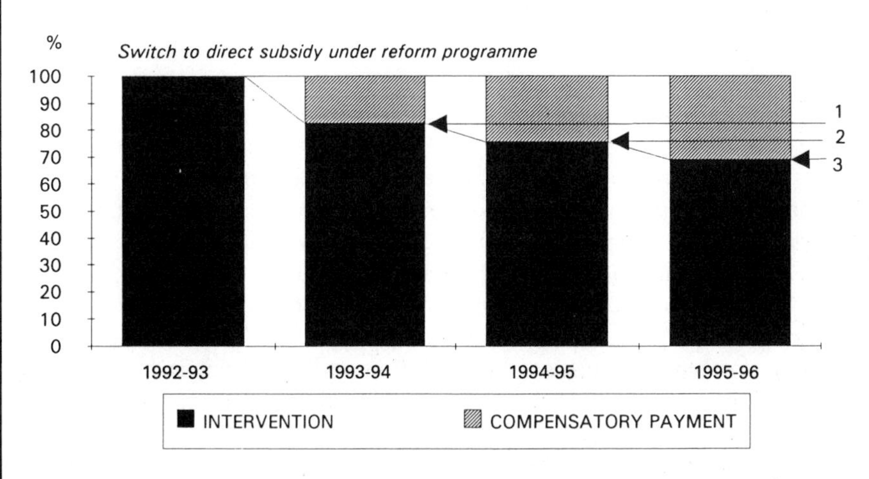

Figure 6.3 Origin of producer return under the new system. Arrows indicate market price levels

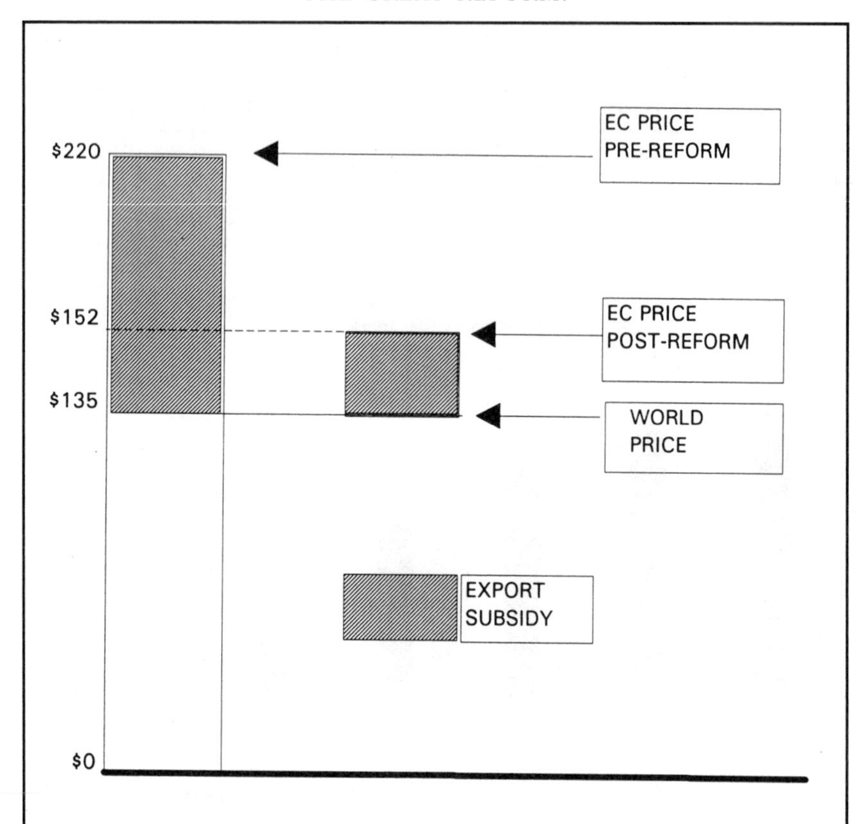

Figure 6.4 Effect of the 1992 reform on the EU's external trade arrangements

The new system involves fundamental change not only of the price levels in the cereal sector, but also of the market price support and external protection structures. Farmers are paid a direct subsidy (crop compensation payment) equal to the difference between the new target price and the original (1992) effective intervention price of 155 ECU/tonne. By the end of the reform period in 1996 this direct payment or 'crop compensation payment' will reach 45 ECU/tonne: the difference between the old intervention price of 155 ECU and the final reformed target price of 110 ECU/tonne. This subsidy is, however, only paid on the basis of the average yield in the period prior to 1992; if a farmer produces more than this basic 'reference yield' he will receive no crop compensation subsidy on the additional amount and will thus receive only the new reduced market price for it.

AGRICULTURAL ENVIRONMENTAL MEASURES AGREED AS PART OF THE 1992 CAP REFORM[13]

As part of the CAP reform agreement, the EU Council of Ministers agreed proposals which marked the beginning of a comprehensive agricultural environment policy. These measures are designed to operate in conjunction with the CAP reform mechanisms.

The measures will pay farmers to farm less intensively by cutting stock numbers, using less fertiliser and pesticides and maintaining what are considered to be environmentally desirable features of the landscape. Most importantly, the 1992 agri-environment measures put the onus on all member governments to make the environmental schemes available to farmers in all twelve countries of the Union. Under earlier agri-environmental legislation it was 'optional' for member governments to apply.

Under this new policy, member state governments are required to draw up five-year zonal programmes for the application of environmental measures, with each country having to divide its territory into zones with environmental policy applications which suit the needs of that area. This effectively means that each country will have its own multi-annual operational programme which would have to be agreed in consultation with the Commission.

Most important are the provisions for the encouragement of extensification. For crops, this will mean a reduction in the use of fertilisers and pesticides designed to reduce both pollution potential as well as output. It will also encourage more extensive forms of production: more rotation of crops, more fallow, more use of traditional grass crops and other similar methods.

For the livestock producer, the environmental legislation supplements the main CAP reform provisions which already encourage a considerable element of extensification – aimed primarily at reducing output of beef, milk and lamb.

Main provisions of environmental measures in the 1992 CAP reform package

There are now a wide range of subsidies for 'agricultural production methods compatible with the requirements of the protection of the environment and the maintenance of the countryside'.

Aids are available under the following Articles of the Environmental Regulation as agreed by the Council of Ministers on 21/22 May 1992:

Article 1

(a) reduction of pollution
(b) extensification of crop farming and stock farming, including the conversion of arable land into extensive grassland

(c) environmental protection
(d) upkeep of abandoned farmland and woodland where necessary for environmental reasons
(e) long-term set-aside
(f) land management for public access
(g) education and training for farmers in environmentally compatible farming

Article 2

(a) reduction of the use of fertilisers and pesticides/maintaining current lower levels/organic farming
(b) more extensive cropping – dedication to extensive grassland
(c) reduction of sheep and cattle per forage hectare
(d) other farming practices compatible with environmental protection – maintenance of countryside, the rearing of livestock and plants in danger of extinction
(e) upkeep of abandoned farmland or woodland
(f) 20-year set-aside
(g) public access management

Article 3

To pay these subsidies, member states will operate multi-annual zonal programmes for a minimum of five years.

Article 4

The level of aids paid under these schemes will be:

- ECU 150/hectare where compensatory payments are already made (i.e. under the main CAP reform price reduction compensation)
- ECU 250/hectare for other annual crops and pasture
- ECU 210 for each livestock unit of flock or herd reduction
- ECU 100 for each LU of endangered breed reared
- ECU 400/hectare of specialised olive groves
- ECU 1000/hectare for citrus fruits
- ECU 700/hectare for other perennial crops and wine
- ECU 250/hectare for upkeep of abandoned land
- ECU 600/hectare for environmental set-aside
- ECU 250/hectare for growing special plants of an environmentally necessary nature

Farmers will also be able to draw a total of ECU 350/hectare if they have given undertakings on fertiliser and crop chemical use reduction or more extensive cropping in combination with 'other farming practices compatible with the requirements of protection of the environment and natural resources'.

THE GATT URUGUAY ROUND AND AGRICULTURE

In November 1992, in Blair House just across the road from the White House in Washington, DC, the European Union and the Americans concluded a deal on agricultural trade with the most far-reaching implications for the world's farm trade ever to be reached. It was the culmination of six years of on-off negotiations, argument and haggling principally between the European Union and the Americans.

The agreement marked the culmination of an ambitious attempt to reduce, even outlaw, subsidising of farmers and the protecting of agricultural markets. It began at the seaside resort of Punta del Este in Uruguay in October 1986 when the signatories, some 109 countries at that time, to the General Agreement on Tariffs and Trade decided not only that there would be a new round of trade talks, the sixth since the Second World War, but also that this round would concentrate heavily on the issue of agricultural trade. Hitherto, agricultural trading had been largely excluded from the other GATT agreements, including most notably the Kennedy and Tokyo Rounds which had immediately preceded this new struggle to liberalise the world's last uncodified sector of international trade.

The Uruguay Round, appropriately enough, began in 1986 which as we have already seen, marked the climax in the continuing struggle of the two major trading forces – the European Union and the United States – for control of world agricultural markets. In that year both the Union and the Americans had spent more than $45 billion in attempting to bribe with export subsidies other countries to buy their wheat, dairy products, beef and other agricultural commodities. Even the Europeans were beginning to realise that this escalating expenditure, this often total waste of taxpayers' money on subsidising farm exports, was increasingly futile.

The beginning of the Uruguay Round coincided with a surge of US Republican antagonism towards other agricultural exporting countries and indeed towards any country deemed to be 'unfairly' subsidising its exports to world markets. Reflecting this extreme and often chauvinistic view, the Administration of then-President Ronald Reagan was being heavily pressured into demanding the total abolition – within ten years – of not only export subsidies on agricultural commodities but also any form of protecting and subsidising of farmers. This so-called trade negotiating 'zero option' aimed for the complete dismantling of agricultural protection by the year 2000. This extreme US position was regarded in Europe as being totally unacceptable. In retrospect, it was a gross tactical error by the Americans.

The US zero option was regarded by all Europeans, liberalisers as well as mercantilists and protectionists, as impractical and, in many areas of the world, undesirable. Negotiations got off to a very bad start; indeed, they did not start constructively at all, but deteriorated into a fruitless transatlantic slanging match. The GATT officials in Geneva spent nearly two years in frustrated attempts to find common ground between the Americans and the Europeans. In adopting the zero option position, the Americans gave the French and other agricultural protectionists in Europe the excuse not to provide the European Union itself with any negotiating mandate on which any subsequent progress could be based.

The so-called Mid-Term Review of progress on the Uruguay Round held in Montreal in the autumn of 1989 demonstrated the largely futile nature of the minimal negotiation which had taken place after more than three years from the launch of the round in 1986.

In the subsequent fifteen months, the GATT Secretary General Arthur Dunkel spent most of his time in permanent orbit between Geneva and Brussels, Brussels and Washington, and back again in an attempt to 'knock heads together' and arrive at some basis for an eventual final negotiation. His 'only' achievement was to persuade the Americans to abandon the zero option. In its place he produced an outline accord which broadly involved a reduction of two-thirds in the level of subsidisation of exports, a 25 per cent cut in subsidies to farmers and an ill-defined commitment to open markets to imports on a volume basis. This prototype agreement formed the basis for what many described as a non-negotiation at what should have been the last session of the GATT Uruguay Round at the Heysel Stadium in Brussels in December 1990. The Heysel meeting ended in deadlock, with the Union and the Americans at loggerheads on even the basic principles, and the Cairns Group of what might be described as liberal agricultural traders standing on the sidelines wringing their hands. The approximate one-third scaling down of protection proposed by Dunkel was still seen by the Europeans as too generous to the Americans.

In the following twelve months, Dunkel spent even more time in the air between the European capitals and Washington, DC. At the end of this time, he had made it plain to the principal protagonists that if there was to be an agreement with any meaning, there had to be a bare minimum of reduction in the budgetary expenditure on agricultural export subsidies, there had to be a minimum reduction in the volume of subsidised agricultural exports, and there had to be a minimum access for agricultural exports to the markets of developed countries.

In December 1991, Dunkel had hammered out his Draft Final Act for what should have been the last negotiating session for the GATT Uruguay Round. The main elements of the Dunkel draft were a 24 per-cent cut in the volume of subsidised exports, a 36 per-cent cut in the expenditure on the subsidising of exports, a 36 per-cent reduction in import barriers through the process of 'tariffication', and a 24 per-cent cut in the level of subsidies paid to farmers in domestic agricultural markets.

This broad outline for the eventual agricultural trade liberalisation agreement immediately became the object of hostility in Europe, particularly from the French and other member-state agricultural organisations. It was condemned as likely to bankrupt farmers, to ruin the agricultural industries and to mean American domination of world agricultural trade.

Despite its very modest features and its pallid reproduction of the original objectives of trade liberalisation enshrined in the Declaration at Punte del Este in 1986, it was quite clear that the 'Dunkel Final Act' would not completely reconcile the opposing views of Brussels and Washington.

The Dunkel draft did, however, provide the basis for continuing negotiation. It also coincided with the agreement some three months later by the European Union of its own reform programme. Although the European officials and politicians were extremely careful not to praise the Dunkel draft in any way publicly, they accepted: (1) that there had to be an agreement on agricultural trade in the context of the Uruguay Round, and (2) that the Dunkel draft must form the basis for the final agreement. It was accepted that there could not be very much further watering down of the Dunkel draft if the agreement was to mean anything at all. The pressing problem for the Europeans, and in particular the tough and wily Agriculture Commissioner Ray MacSharry, was to make the potential GATT agreement fit the recent modification of the Union's own agriculture policy – rather than the reverse.

It was for this reason that the EU negotiators continued publicly to argue that Dunkel's Draft Final Act was completely unacceptable. Behind the scenes, however, continuous and feverish negotiation went on between Brussels and Washington in order to find ways of modifying Dunkel in order to construct an agreement which would be acceptable to the Europeans and in particular to the French Government. The negotiations went on throughout the summer of 1992 at summit meetings of heads of government, but the real, tough bargaining went on behind closed doors in Brussels and Washington and between the experts of the European Commission's Agricultural Directorate and the US Government's Department of Agriculture. These behind-the-scenes negotiations culminated in the pivotal Blair House meeting and agreement on agricultural trade in Washington in November 1992.

Although prepared to compromise, the Americans were determined to obtain specific concrete advantages from what they knew would have to be further concessions to the Europeans on the broad aspects of agricultural trade arrangements. A major preoccupation of the USDA at that time was the increasing expansion of EU oilseed production, which had not only displaced US soyabean and other oilseed exports to the European Union domestic market but was beginning, as in other products, to become a potential threat to American exports elsewhere in the world. The Americans were adamant that they must have limitations on EU oilseed production as a quid pro quo for a more flexible position on the broader agricultural trade elements of the GATT Uruguay Round.

They therefore bartered a cut in subsidised agricultural exports of only 21 per cent (in place of Dunkel's original 24 per cent) and a more liberal view on the dismantling of support for farmers, for a European Union agreement to limit its oilseed production to the 1989/90 level of output and to subject oilseeds to the same set-aside requirement as were already envisaged by the Union for its cereal crops. After much under-cover negotiation between the European Commission and the member-state governments, the Union finally agreed both the oilseed production limitation and the new version of the Draft Final Act at the Blair House meeting in November 1992.

This bilateral arrangement was subsequently agreed at the level of the Council of Ministers in the European Union, and it was agreed by the Europeans that the agricultural trade portion of the Blair House agreement should form the basis of the eventual multi-lateral agreement on agricultural trade which would complete the Uruguay Round negotiation.

Unfortunately for the smooth completion of the round, both the United States and the French governments changed in the months immediately following conclusion of Blair House. These changes came close to de-railing the whole GATT negotiation.

The new US Government of Democrat President Bill Clinton, unversed in the intricacies of foreign affairs, and determined to adopt a different position from its Republican predecessor, issued a series of inflammatory statements on the GATT within hours of taking office which were eagerly retailed in Paris and other European capitals and formed the beginnings of a reaction against the bilateral agreement on farm trade reached in Washington in November 1992. This was to be the cause of further watering-down of the eventual multilateral GATT agreement on farm trade. It meant further loss to the Americans and gains to the Europeans in the final agreement.

The Washington 1992 accord was further eroded with the defeat of the Socialist Government of François Mitterrand in the April 1993 French parliamentary elections and its replacement with a right-wing alliance determined to put 'France first' and nowhere more so than in the agricultural sphere.

The situation was made worse by the style and content of the French election campaign leading up to the April parliamentary elections, with considerable rivalry between the right-wing Gaullist faction, led by Jacques Chirac, and the more moderate group RPR led by Edouard Balladur. The Chirac faction rejected the Blair House agreement out of hand, and throughout the election campaign maintained that the American intention in the GATT was the destruction of French and European agriculture, and the domination of world agricultural trade. In order to maintain its position and combat the Gaullist pressures, the Balladur faction unfortunately gave hostages to fortune in the form of a commitment to reject Blair House.

Following its election victory, the new French government of Edouard Balladur had the difficult job of going back from this somewhat extreme position, knowing full well that a full rejection of a GATT agreement on agriculture was out of the question. The French Government was under heavy pressure from its own industrial and business sector supporters, who pointed out that though a GATT agreement could be marginally harmful for French agriculture, it was essential to the industrial export ambitions of France. The Balladur Government therefore developed the idea of 'redefining' the terms of the November 1992 Washington accord on agricultural trade.

Statements by Balladur and his aides that France intended effectively to renegotiate the Blair House agreement met a hostile reaction in Washington and a stand-off on the agricultural trade negotiations for the three midsummer months of 1993. At this point there was general disappointment throughout the agricultural world and a deep fear that the new division between Washington and Paris on the agricultural issue would mean the fruitless end of the GATT Uruguay Round.

What made matters worse was that the European Commission itself had changed in the period since November 1992 and the undoubtedly remarkable negotiating talents of Ray MacSharry were no longer available to the Union. Similarly, the team in Washington lacked experience and did not appear to have the same enthusiasm for an eventual agreement as the Bush team had shown during the autumn of the previous year.

None the less, there were extensive private meetings between European Commission and US Department of Agriculture officials. The Europeans stressed that something had to be brought out of the hat to satisfy the French Government and the French farm unions if there were to be an agreement. In the early autumn of 1993 it emerged that the most important problem for the Union, as seen through French eyes, was the dissipation of the stocks particularly of wheat and other cereals but also beef, which had accumulated as a result of the mismanagement of the agriculture policy in the 1980s and which the new reforms of May 1992 could not possibly dissipate before well into what would be the new period for the application of the GATT Uruguay Round conditions (1995 to approximately 1999/2000).

The Europeans wanted a concession built into the eventual agreement which would allow them to dump some 25 million tonnes of wheat and half a million tonnes of beef on international markets with subsidies – outside the terms and limits of the eventual GATT agreement. This request was rejected out of hand by Washington; it was something which could not be accepted by an administration dedicated to being 'tougher with foreigners' than its Republican predecessor. However, in the classic manner of international negotiations, the officials of both sides managed to dream up a formula which would effectively result in a breathing space for the Union on surplus disposal, while at the same time not giving the appearance to normal, but non-specialist, newspaper readers that any concession had actually been made.

This winning formula carried the remarkable label of 'front-end loading'. It consisted of phasing the reduction in subsidised exports over the now bilaterally agreed GATT period of 1995–2000 so that the bulk of the reduction took place not in the first two and a half to three years of this six-year agreement, but in the latter years. In other words, the amount of subsidised exports which the Union would be able to send to world markets would not reflect an equal annual reduction in exports as had originally been envisaged, but would allow the level of subsidised exports in the first years of the agreement to remain more or less unchanged compared with the non-GATT situation. The bulk of the reduction under this new agreement would take place in the last two and one half to three years of the implementation of the Uruguay Round agreement.

This front-end loading stratagem was the centrepiece of what came to be known as the Blair House II or 'Breydel' agreement reached at the Commission's Breydel headquarters in Brussels in December 1993. The conclusion of this bilateral accord between the world's two major agricultural traders in fact opened the way to the final agreement in the Uruguay Round of the GATT in Geneva in December 1993. Although there were still other issues to be settled, such as intellectual property – particularly the problem of freedom to trade in films and other artistic commodities – the Breydel agreement, as it came to be known, undoubtedly broke the log-jam and paved the way to the final Uruguay Round outline agreement in Geneva and the final ministerial accord in Marrakesh in April 1994.

To the detached observer, it may seem strange that the French and other European farm unions took such exception to the very modest terms of the Blair House I agreement when it would appear that their governments had agreed reform of the domestic EU agriculture policy and adjustments of the external protection measures in 1992, which, when looked at objectively, would have effects on EU production and exports very similar to the terms of the US–EU bilateral pre-GATT agreement. It was however a major contention, and still is, among French agricultural interests, and indeed among objective analysts of the EU agricultural scene, that the 1992 CAP reforms do not go far enough to meet the terms of what is now the GATT Uruguay Round Agreement. The French farm-lobby view, based on this assumption, is therefore quite logical: further reforms beyond the 1992 programme will be needed to meet the terms of the GATT agreement.

This school of thought[14] believes that further domestic reforms will be needed for the EU to meet the terms of the GATT agreement. The important point at issue is how far the CAP reforms will discourage farmers

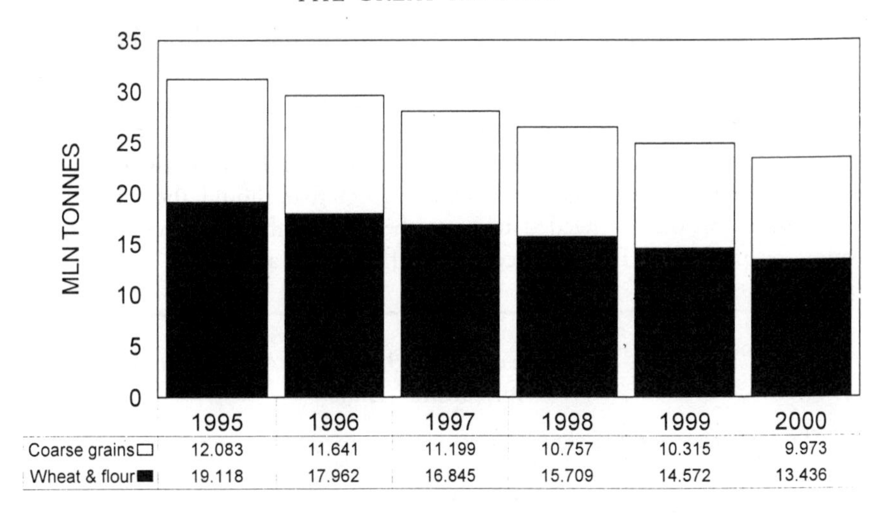

	1995	1996	1997	1998	1999	2000
Coarse grains □	12.083	11.641	11.199	10.757	10.315	9.973
Wheat & flour ■	19.118	17.962	16.845	15.709	14.572	13.436

Figure 6.5 The EU's GATT export limits for cereals.

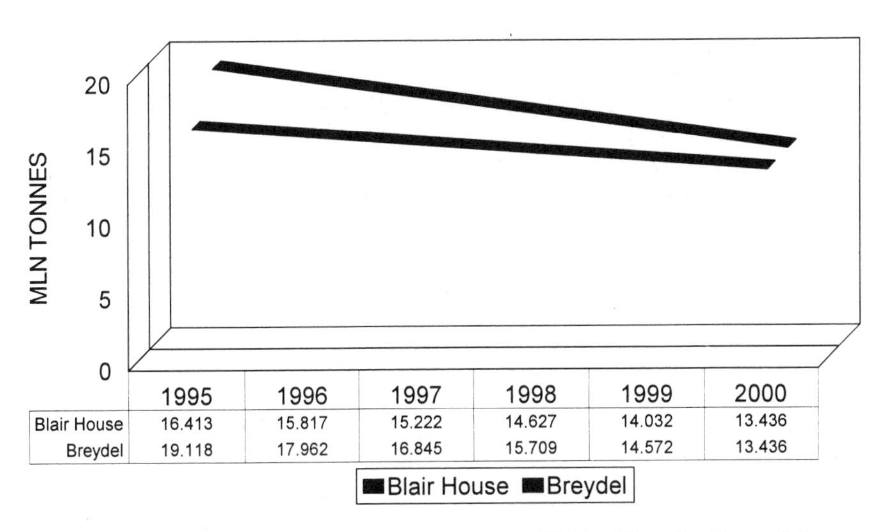

	1995	1996	1997	1998	1999	2000
Blair House	16.413	15.817	15.222	14.627	14.032	13.436
Breydel	19.118	17.962	16.845	15.709	14.572	13.436

Figure 6.6 Maximum EU wheat and flour exports – 'front loading gain'

from maximising their production – in other words, exploiting the possible productivity-increasing innovations which may or may not be available to them during the period between now and the end of the century.

The French position remains that productivity will increase faster, that the total level of EU output in the 1995–2000 period will exceed the limits laid down in the Uruguay Round Agreement and that the Union will therefore be forced to introduce further limitations on EU production in order to meet the GATT requirements.

The Commission position,[15] which has been accepted by other governments, is that productivity increases will not be large enough to increase production beyond the limits set in the Breydel/GATT agreement. The French Government was so convinced of the incompatibility of the CAP reforms with the GATT that it insisted at a special meeting of Foreign and Agriculture Ministers in November 1993 that should there be further limitations on output needed to meet the GATT, then extra compensation must be paid from EU resources to the farmers as a result.

The Main Elements of the 1993 GATT Agreement on Agricultural Trade

The most important elements of the agreement applying to agriculture are: a 36 per-cent cut in budgetary expenditure on export subsidies, a 21 per-cent reduction in the volume of subsidised exports and a 20 per-cent scaling-down of domestic supports – on the basis of the average levels operating in the period 1986–90 for exports and 1986–88 for internal support, with the reductions to be phased over the period 1995–2000 (2000–2001 marketing year).

The new rules applied from July 1995 and are to be implemented over the following six crop years to 2000. The plan recommends that import levies, export subsidies and non-tariff barriers should be reduced by the mechanisms of 'tariffication' – the conversion of all barriers to

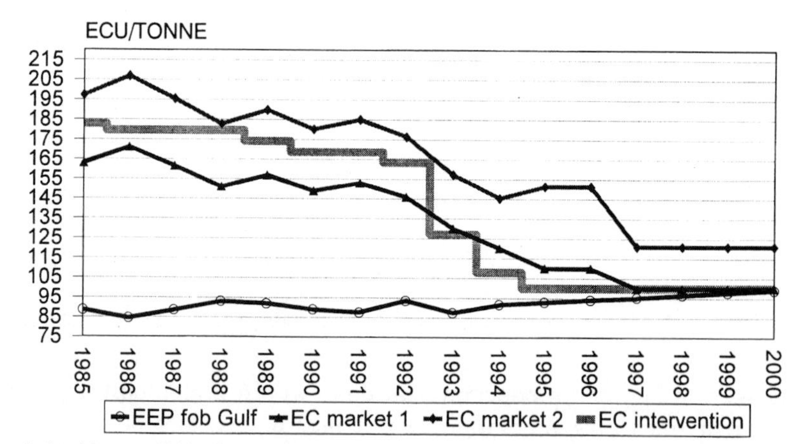

Figure 6.7 Probable impact of reform and GATT on the relationship between EU price levels and the world market – narrowing the EU/world price 'wedge'

a tariff equivalent which then forms the basis for reductions. The agreement says that this process must result in a minimum reduction in the tariff of any individual product of 15 per cent. Similarly, there must be a minimum increase in import access of any products of 3 per cent, rising to 5 per cent of domestic consumption – on the basis of the average of 1986–88 – by the end of the 1996–2000 period.

'Tariffication'

Setting a tariff basis for world agricultural trade is seen by the United States and other exporting-country negotiators as not only a means of putting a ceiling on agricultural support and protection, but as providing the means for applying effective pressure for further reduction. Tariffs are seen by their proponents as having several major advantages over other, less clear mechanisms for protecting and supporting agriculture. They have the advantage of being: (1) transparent, (2) linking domestic price levels in protecting countries to the world price, (3) allowing the revenue from tariffs used to protect one sector to be switched to assist others, and (4) they minimise the protection of domestic industries. Tariffs, it is said,[16] are measurable and comparable at the international level.

International setting of tariff levels means that a relationship is established between the movement of world prices and domestic support prices. The main elements of the tariffication proposal are: the inclusion of the four main areas of import obstruction and protection – import levies and other barriers, export competition, domestic support and sanitary and phytosanitary non-tariff barriers – in an overall protection-reduction programme. All of these components should be converted into tariff equivalents (TEs), initial limits established and a timetable then set for their reduction and eventual elimination. The practical assessment of the TE will be on the basis of the observed difference between internal and international prices in an agreed period.

7

IMPACT OF CAP REFORM AND THE GATT URUGUAY ROUND AGREEMENT ON AGRICULTURAL TRADE

What is likely to happen as a result of the European Union's policy innovations of the early 1990s and the application of the 1993 GATT agreement from 1995 onwards? First the facts. The basis for the GATT Uruguay Round agreement on agricultural trade was the Draft Final Act published in December 1991 (the 'Dunkel Plan'), as modified by the Blair House agreement of November 1992, the US–EC bilateral negotiations of November/December 1993 and the final accord reached at Geneva in the second half of December 1993 (see Chapter 6). The full rectified Uruguay Round agreement was agreed in Marrakesh in April 1994.

As the evidence presented in Chapter 6 indicates, the EU reform programme is concentrated upon the cereals sector, because of its central role in agricultural production, incomes and exports; it will have a much smaller impact on the livestock sectors. Much the same can be said of the GATT agreement itself: the central battle and the terms of the eventual peace treaty centred upon world trade in grain. For this reason the assessment of both CAP reform and GATT in this chapter concentrate upon the effect on cereal production and trade.

By limiting the amount of grain which the Union may export with subsidies, the GATT agreement is likely to impose further constraints on the EU arable sector beyond those imposed by the CAP reforms. There is, however, considerable controversy over just how much further reduction in output would be needed to meet the agreement to reduce subsidised exports by 21 per cent of the 1986–89 average level over the 1993–9 period. The net reduction of wheat exports would have to be over 40 per cent from the 1992–93 level of 21–22 million tonnes to reduce them to the 12-13 million tonnes indicated by the Blair House condition.

There are a number of key factors to be considered, the most important of which are:

1 the extent to which productivity will or will not increase;
2 the effectiveness of EU production limitation policies;

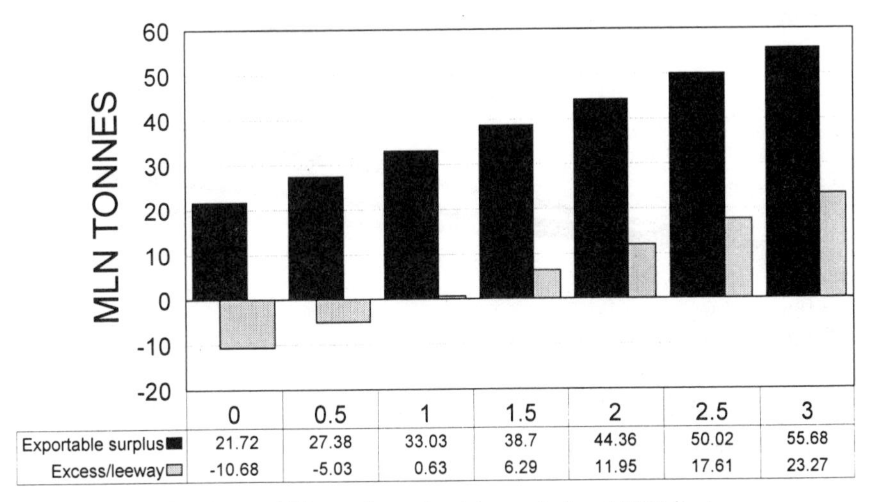

Figure 7.1 EU cereal productivity and the GATT limit

3 the degree to which the world price will rise (it is generally assumed by most analysts that the GATT agreement will result in a general rise in agricultural commodity prices) and the EU domestic market price will fall, thus reducing the export subsidy requirement;
4 the ability or otherwise of EU cereal growers to produce profitably at close to the higher, post-GATT world price.

There are two possible 'main scenarios'. If productivity increases at the historical level of 1.5 to 2 per cent, the gap between the world and the EU market price remains large and EU producers do not reduce their production costs, then the Union will have problems in meeting the GATT reduction target. If, on the other hand, the world price rises substantially and EU farmers become more cost-efficient, then there is no great problem: EU farmers will be able to compete at the new world price and be able to export a significant proportion or all of their output without subsidies.

There is clear evidence that the best of EU cereal growers – those responsible for the bulk of the output – are capable of competing in the world market without export subventions or any other form of subsidy. A US Department of Agriculture study[1] showed quite clearly that cereal growers' production costs in the United Kingdom and France are little different from those of their American counterparts. Purchasing parity advantages of the US farmer would be likely to be counterbalanced by lower transport costs from Europe to key export markets. Evidence from EU analyses[2] is increasingly supporting this conclusion too.

Assessments of the likely effect vary from the optimistic view that production will decline significantly due to the CAP reforms and thus the subsidised

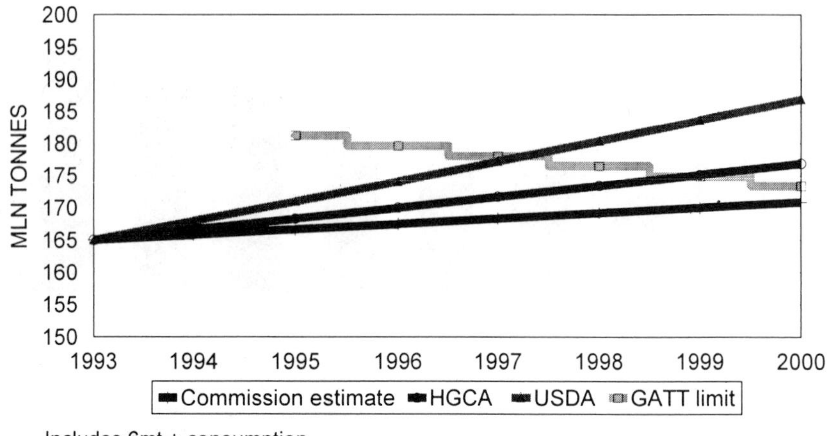

Figure 7.2 EU cereal production in relation to the GATT limit – alternative scenarios?

export reduction commitment will be met without further reductions in cereal area,[3] to the most pessimistic view that production will continue to increase, the world price will not rise and that increased output would not be able to be exported without subsidy.[4]

This latter eventuality would bring the EU agriculture industry into conflict not only with the GATT, but also with the Union's own agricultural budgetary limitations: further production restrictions would be inevitable.

If the Union cannot reduce its price to the world level and thus eliminate export subsidies, it has to reduce its subsidised exports of wheat by some 7.5 million tonnes on the 1992 level by 1999 and its total subsidised cereal export by 13 to 14 million tonnes. The EU Commission's view is that the CAP reforms will automatically result in an export requirement reduction sufficient to meet the needs of the GATT agreement.

Most likely (4 above) is that EU crop productivity will increase modestly and production will tend to increase after the initial effect of set-aside has diminished (post-1995–96). The world price of cereals is likely to rise, not because of any significant reduction in world exports, but because of the significant reduction in export subsidies by both the Union and the Americans.

The Commission, on the other hand, believes[5] that there will be no significant increases in productivity in the cereal sector and that farmers will not produce beyond the quantity prescribed by the crop compensation payments (the regional historical average yield). The Commission also expects a total increase in EU domestic cereal consumption over the period to 1999 of 8–12 million tonnes. On this basis, it expects that 'the exportable surplus will remain within the limits authorised in the draft (Blair House) agreement and be compatible with CAP reform'.

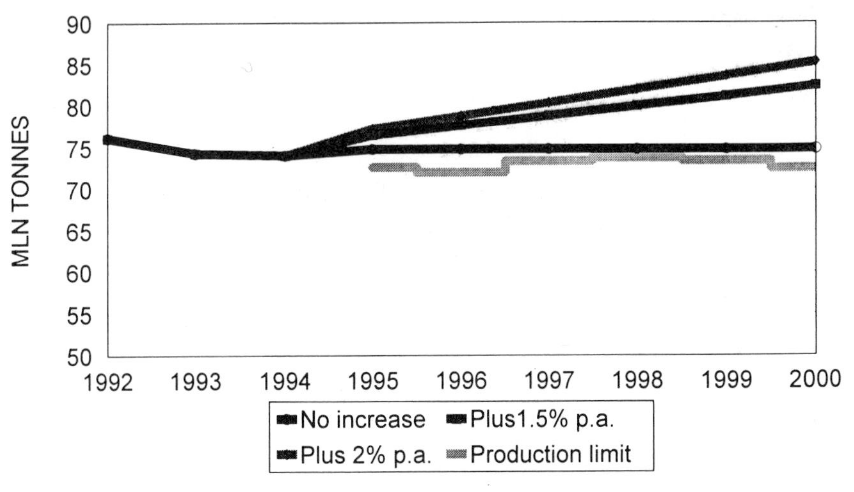

Figure 7.3 EU wheat production in relation to GATT limit?

Note: Production limit = consumption + GATT export limit

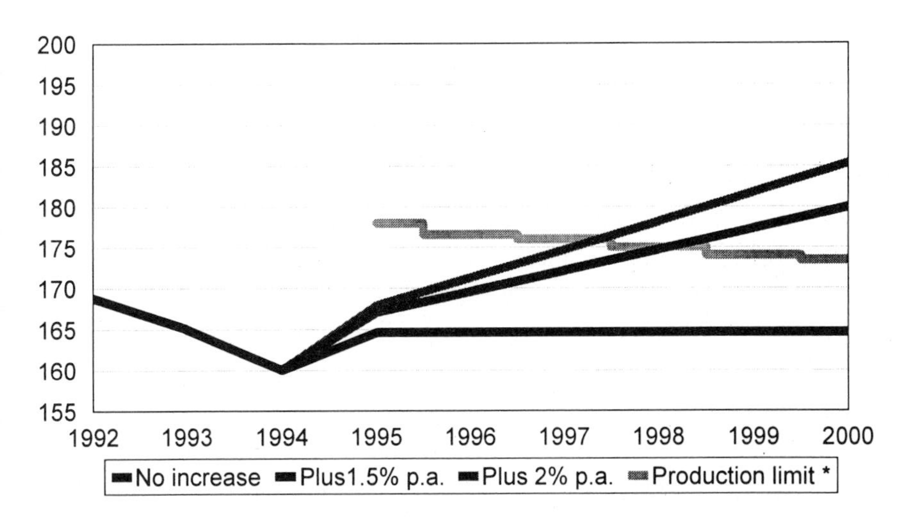

Figure 7.4 EU possible total cereal production in relation to GATT limit?

Note: Production limit = consumption + GATT export limit

The Commission argues that the evolution of production will depend on three variables: production, imports and consumption. It stresses that the estimates which it made in July 1992, subsequent to the reform agreement, were based on a stable average cereal yield of 4.80 tonnes/hectare (1992–93: 4.66 tonnes/hectare). The Commission argues against analyses which postulate large increases in productivity.

Independent analysts suggest[6] at least five possible flaws in the Commission's argument:

1 That productivity will be greater than the Commission expects. Productivity levels are likely to rise rather than fall in the latter half of the 1990s. Even assuming, however, a very modest 1.5 per cent a year, this results in a total cereal production of 172.4 million tonnes by 1999. This is 13.4 million tonnes greater than the limit indicated by Blair House (after allowing for a 6-million-tonne increase in domestic consumption).
2 They question the assumption that there is no incentive for cereal growers to increase their average yields under the new compensatory subsidy system; this assumes that farmers will not seek to maximise profit by producing more than is covered by the subsidy. It is, however, likely that in a situation of probably falling costs, it will still pay efficient producers to grow marginal tonnes beyond their reference quantity (on which compensation is paid) – even at the post-1996 intervention supported level of 100 ECU/tonne.
3 No allowance is made in the Commission's estimates for the initial boost in output on the 85 per cent of land (or 90 per cent which it became in 1995–96) that remains in production after 15 per cent (10 per cent in 1995–96) is set-aside and the resources previously used on the full area concentrated upon the reduced area.
4 No allowance is made for the inevitable 'slippage' effect on the effective set-aside. While the Commission initially adjusted its average set-aside production reduction impact to 9.5 per cent to allow for the under 92-tonne-a-year producers who are exempt from set-aside, the probable effect of set-aside on production will be even lower – a net 7.5 per cent is suggested by comparison with the US[7] – or about half the nominal figure.
5 Lastly, the Commission's estimates of increased feed use are questionable. Grain-fed livestock production is stabilising and, with limits on export refunds on pig and poultry products, demand is expected to fall. In addition, as the price of cereals falls, so too will the price of 'substitutes' (cheap imported carbohydrate materials such as manioc, citrus pulps, sugar beet pulp and so on). The scope for replacement with cereals will therefore be limited. A projected increase of half the Commissions's figure (6 million tonnes) is thought to be more reasonable.

How will the agreement fit with the CAP reforms (see Chapter 6) already implemented by the European Union in 1993?

EFFECTIVENESS OF SET-ASIDE?

One of the most contentious issues is likely to be the impact of setting aside land from production – how much overall reduction in output will this achieve?

The EU Commission believes that its initial nominal 15 per-cent arable setting-aside of arable crop land will result in a 9+ per-cent effective set-aside of cereal-producing land, after the effects of the exemption of the less than 92-tonne-a-year producers is allowed for. The tendency of farmers to concentrate their arable set-aside on less profitable arable crops than cereals, combined with the increasing tendency for farmers to concentrate the same amount of resources on the area of land that does remain in production, plus the fertility-enhancing impact of set-aside land itself, will mean that the effect on cereals production will be substantially less than the Commission's expected figure. In the United Kingdom major East Anglian grain-producing area for example, the actual setting-aside of wheat growing land in the first year of CAP reform was no more than 4 per cent.[8] The Commission's assumptions about the effectiveness of set-aside are likely to be proved to be highly over-optimistic when the effect is averaged over the 1993–96 reform period.

The problem for the Commission and the EU is that set-aside is likely to become less, rather than more, effective with each year that passes. Seen in both a GATT and an EU budget context, the biggest danger in the development of the set-aside policy is the scope which it is giving to maximise the area and production of wheat, at the expense of other grains and oilseed crops. Because of the greater profitability of wheat compared with any other crop covered by the set-aside requirement, farmers will tend to increase their wheat area to the maximum: an important distortionary effect of the new policy.

An on-farm survey of EU farmers' response to set-aside by a group of economists during 1993[9] shows quite clearly that farmers are likely to maximise wheat area within the set-aside limit – because it is most profitable to do so, to select high yielding varieties, to maintain wheat on the best land and to seek to minimise the proportion of their best land that is actually set-aside.

The three main conclusions of the survey were:

1 Because participating farmers will tend to idle their least productive land, average yield will be higher on the land remaining in production. Introduction of a non-rotational set-aside option in 1994 further contributed to slippage.[10] At a 15 per-cent SA rate, 90 per cent of eligible area would be idled at one time during the six-year period. The additional non-rotational set-aside will allow producers to plant their most productive land every year.
2 Intensification of production on land remaining in production: this will allow farmers to maintain or increase yields even if chemical input use is adjusted in response to lower prices.
3 Yields will increase due to set-aside itself: idling for one year increases the productivity of the land through retention of higher levels of soil moisture and nutrients.

The survey indicated that there is likely to be a marked move to concentrate upon wheat at the expense of other crops. This has subsequently been proved true by events: Commission figures for crop area show a continuous trend of relatively undiminished wheat area, with oilseed and protein crops bearing the brunt of the set-aside requirement.[11] A lower effective set-aside rate for wheat than for other arable crops is resulting in higher average grain yields, as common wheat makes up an increased part of the total grain area.

Average wheat yields themselves can be expected to increase, as producers will grow fewer lower-yielding second and third wheats. Wheat yield slippage in the EU can be expected to be greater than in the United States because wheat is grown on the best land – the reverse of the US situation.

EFFECT OF THE GATT AGREEMENT ON US CEREAL PRODUCTION AND EXPORTS

Since the major influence on the world's grain – and food markets generally – is US production and exports, it is essential to assess the impact of the GATT on US agriculture. In fact, the effect of a GATT Uruguay Round agreement on US cereals policy and hence on the level of production is likely to be small. The reason for this is simple: the United States has made significant reductions in the support of its cereals sector in the period 1989–92. Since it was already operating at a much lower rate of support price, the United States has to make much less adjustment than the EU. Indeed, its role as the progenitor of the agricultural trade reform initiative in the Uruguay Round will be justified by its likely expansion of production and exports as a result of the GATT impact. US reductions in support in the 1986–95 period amount very nearly to the total amount of reduction in domestic support required under the terms of the Uruguay Round accord.[12]

The US support level for the main grains is very close to the world price; if the world price rises as a result of the GATT agreement, then the gap between the world price and the US support level is reduced and the need for support diminished. The 1994 guarantee level (Target Price) for wheat in the United States was approximately $146.97 a tonne; if the world price rises in the period of the GATT agreement from the 1994 level of $140 a tonne to $147–150/tonne – as it did in the 1994–95 period, then the need for support disappears. So too does the need to subsidise exports: without export subsidies there will be no GATT limitations on the quantity that the United States might export.

The agreement involves reductions in both the level of budgetary expenditure on export subsidies and the volume of subsidised exports. It is, however, likely that these would be sufficiently reduced automatically. This is because as the levels of subsidised exports of the European Union are reduced by the terms of the agreement, world prices are likely to rise towards the US domestic support level. As the EU reduces its subsidy, so the US subsidy will fall and will possibly be eliminated by the end of the GATT period.

A USDA assessment[13] of the effect of the GATT agreement suggests that wheat exports will increase by 10–12 per cent, maize exports by 5–6 per cent and barley exports by 3–8 per cent. The 1990 Farm Bill[14] initially resulted in an overall reduction in the level of support to US agriculture of close to 5 per cent. The terms of the Act are likely to result in further reductions in support in the period to 1995–96. The 1995 US Farm Bill will add to the expansionary nature of American agriculture in the later 1990s by minimising set-aside requirements. Under the 1990 legislation for cereals, both reductions in area and in the level of subsidy were envisaged. Since this legislation was largely designed to fit into the world market situation created by an expected Uruguay Round agreement, it is not surprising that little further adjustment is likely to be needed in US cereals policy to comply with the 1993–94 GATT agreement.

As the world price rises, so the level of US Government export subsidy can be expected to decline, but the return to the producer will remain largely unchanged as the increased world price compensates for reduced subsidy. The reduction in subsidy commitment will allow the US Government to relax its restriction of planted area, since the purpose of such restrictions (the Area Reduction Programme – set-aside) is solely to limit the budgetary expenditure on export subsidies and other forms of support.

Since it is not price, but rather official acreage limitations, which prevent US cereal production from expanding, US cereal output can be expected to increase in the later 1990s. Substantial reduction and eventual elimination of planting restrictions is likely to be the major effect of a GATT agreement on the US cereal sector; it is also likely to be the major influence on world prices.[15]

The main influences on the future level of US wheat and coarse grain production are therefore likely to be (1) the elimination of set-aside, (2) world price levels, and (3) productivity increases. Since it can be assumed that set-aside will be eliminated – in stages, in line with the application of the GATT agreement – and that the world price will rise close to or above current US support levels and will not act as a disincentive to production, the main influences on future US output will be the removal of planting restrictions and any increase in productivity.

Productivity increases in the United States are likely to be more modest than in Europe, because there is less unexploited productive capacity in North American agriculture than there is on most EU farms. It is therefore justifiable to expect a modest annual average increase in productivity of 0.5 per cent a year through the later 1990s. Combined with the gradual removal of set-aside and only small increases in domestic consumption, this is likely to mean that US production and exports of wheat and coarse grain will rise from 1994–95 onwards to return to the levels of the mid-1980s by the end of the decade. On this basis, it is likely that US wheat production will rise by 16 per cent from the 1992 level of 56.6 million tonnes to 65.6 million by 2000. Exports of wheat can be expected to increase from the 1992 level of 24.5 to 34 million tonnes – an increase of 39 per cent. The respective figures for coarse grains are: production rising 18 per cent to 266 million tonnes (on the 1989–91 average; 1992 was an exceptional harvest) by 2000 and exports increasing by 43 per cent to 79 million tonnes.

Net US exports of cereals could be expected to increase by 163,000 tonnes for each 1 per-cent cut in target price and to decrease by 1.44 million tonnes for each 1 per-cent increase in the acreage reduction requirement.

THE GATT AND THE WORLD MARKET

A major factor determining the world prices of cereals and other major farm commodities in the period 1994–2000 is the level of production of and prices for these commodities in the European Union. In the case of dairy products, beef and sugar this is because of the predominant share of the EU in world exports; in the case of cereals it is because the export subsidy policy pursued by the Union will also affect closely the extent to which the United States will subsidise its agricultural exports. While the EU has only an 18 per-cent share of the world wheat market, for example, its past tendency to subsidise these exports heavily has meant that the United States has also subsidised the bulk of its 35–40 per cent share of the world market – resulting in an important combined depressive effect on the world market. If the Union reduces its cereal-exports subsidies, so too will the United States: world prices will consequently rise. Since the United States and the Union between them are responsible for more than 50 per cent of the world trade in wheat, the substantial reduction and eventual elimination of their export subsidies will raise world prices.[16]

The GATT arrangements will inevitably lead to limitations on the EU's subsidised cereal exports after 1995. Given the expected fall in production costs resulting from productivity increases, plus also the fall in the EU price level, there is nothing to prevent the Union from exporting a significant proportion of its output without subsidy. It is assumed therefore that an important proportion of the EU grain output will, by the late 1990s, be produced at close to world price levels (this also assumes that the world price will rise on the strength of a Uruguay Round agreement).

This is why most EU export estimates for what might be called the 'GATT agreement' period (1995–2000) are for several commodities – wheat in particular – greater than would be expected to be permissible under the agreement to cut the volume of subsidised exports by 21 per cent over the 1994–99 period.[17]

It is also assumed that there will be productivity increases in all sectors of the EU agricultural economy, but most particularly in arable cropping and dairy production. Considerable differences in the effect of the EU's 1992 reforms on the cropping and livestock sectors from the effects on cereals and other crop commodities have to be borne in mind. This is because the CAP reform price adjustments in the livestock sector form a very much smaller proportion of the total producer return and because the direct compensation to farmers would more than cover the resulting loss in income to the producer.

EU production of wheat declined by approximately 4+ per cent from just over 80 million tonnes in 1992 to close to 76.5 million tonnes in 1993, as a result of the imposition of set-aside. By 1995, however, production had increased again to close to 80 million tonnes. There were to reasons for this: continuing increases in yield, but more importantly, the tendency of farmers to minimise wheat set-aside by concentrating their set-aside liability on other eligible crops such as oilseeds. Output per hectare can be expected to continue to increase by at least 1.5 per cent a year, with EU output of wheat recovering to approximately 85 million tonnes, significantly above the 1992 level, by 2000.

The increase in supply could be to some extent counterbalanced by an increase in consumption stimulated by lower market prices: from 59 million tonnes in 1992 to 65 million tonnes, a rise of just over 10 per cent.[18]

The EU's total grain export potential is likely to fall from a peak of more than 32 million tonnes in 1992 to 19.6 million tonnes in 1994, a reduction of 30 per cent. This reduction will be due exclusively to the impact of set-aside, but continuing increases in productivity (1.5 per cent per annum) will result in a rise in export possibility to close to 23 million tonnes by the end of the century. This means, therefore, that the Union's likely reduction in wheat exports would be a net 18 per cent in 2000 compared with 1992.

The world market price, on the assumption of the phasing out of export subsidies by both the European Union and the United States, can be expected to rise by at least 1 per cent a year on the 1990-92 average of US$140/tonnes to $150 plus by 2000 – an increase of 7 plus per cent.[19]

EU SUPPORT PRICES AND THE MARKET REALITY

The crucial question in assessing the effectiveness of reform and the impact of the GATT is the profitability of EU cereal production in relation to likely future levels of market price. The 1992 reforms are likely to result in an EU internal market price somewhere in the range of 95 to 110 ('green') ECU/tonne by the 1996–97 cereal marketing year. Assuming that the GATT agreement raises the world price by 6 to 10 per cent on the 1989–93 average, then it is likely that the Union's cereal producers would have to expect an average price somewhere in the 85–95 (green) ECU/tonne by the end of the decade. This assumes that the Union will seek to avoid the GATT export limitation by eliminating export subsidies through reducing the EU price level to equal the 'new' post-GATT world price. How many cereal growers would survive at these price levels and what would their aggregate level of production be?

The answer to these closely linked questions can only be answered by postulating several alternative lines of development. The differences between such 'scenarios' would arise mainly from varying levels of productivity. It would be reasonable to postulate three main alternatives: (1) that there will be no increase in productivity, (2) that productivity will increase at a modest 1.5 per cent between now and the millennium, and (3) that it could increase by an outside 2 per cent per annum in the same period.

The important question is: will production increase, decline or remain the same under these three most likely scenarios?

The most important conclusion is that, even with the compensatory subsidy system introduced in 1992, efficient EU cereal growers have everything to gain by maximising output and a great deal to lose by not doing so. Under the 1992 CAP reform programme, efficient cereal growers will make an easy net profit of around 200 ECU per hectare if they merely stick at a modest yield level of 6.5 to 7 tonnes per hectare – even at the likely post-CAP reform (1996–97) wheat price of around 100 ECU/tonne. An important proviso, of course, is that the Area Compensation payments are maintained at their full level to the end of the decade.

The gains from maximising yield, rather than merely sustaining output at the Reference Yield level, are substantial. The difference in profitability between production at the Reference Yield level and production at the quite feasible 8.5 to 9.5 tonnes per hectare range is close to 100 per cent. These calculations are based on the costs of efficient cereal growers, and assume that shifting from an average level of output to maximising output is likely to involve a significant rise in variable costs somewhere around the 7.5 tonnes per hectare point of output. The gains from productivity in cereal growing are also likely to be substantial: an increase of 1.5 per cent a year in the period between now and the end of the decade is likely under these farm conditions to result in an increase in profitability of more than 20 per cent – a further indication of why farmers are likely to maximise output rather than merely being content to stabilise output at their Reference Yield levels.[20]

The most efficient producers would be able to remain in production, even if there were no compensation payments. This is clearly more likely when the productivity increases are taken into account.

The implications of a possible post-reform, post-GATT adjustment of the EU market price of wheat to the possible world price level (in order to avoid GATT-subsidised export limitations?) may also be deduced from these calculations. The most obvious conclusion is that only the most efficient could survive at this price without compensation, while the compensation system makes it possible for producers to achieve a net income of close to the current (1993–94) efficient average specialist arable farm level of around 220 ECU/ha at a yield of 8+ tonnes/ha. With a 10 per-cent increase in productivity that level of income can be achieved at a yield of around 7.75 tonnes/ha.

These figures are based on mid-1990s cost levels, and it is therefore possible that further increases in efficiency are likely. Most farm management experts believe that considerable savings on fixed costs are still possible even on the most efficient of European arable farms. Most farms still carry too much labour and are over-capitalised in the sense of having excess mechanisation capacity. Average fixed costs can still be substantially cut by reducing labour and the pooling of machinery – either through syndicates or the greater use of contractors. It is expected that the 'shock' effect of the CAP reforms and the GATT will make farmers look more closely at their cost structure and that therefore there will be further reductions in costs. This would have the important effect of sustaining profitability at lower prices – even at the lower yield levels.

Taken together, all these factors would suggest that the EU's cereal sector is more likely to expand its output – on the post-1992, set-aside adjusted level, than to contract or even stabilise.

COMPATIBILITY WITH GATT?

How do the CAP reforms fit in with the new GATT arrangement? In other words, will further modification of the CAP be necessary? A 29 per-cent cut in the EU support price for cereals should involve a 60–80 per-cent reduction, depending on world market price levels, in the level of export subsidy needed to maintain the competitiveness of EU cereals on the international market. Reducing the EU internal cereal support price reduces the (normal) large gap between the EU market price levels and the world price. It follows therefore that the export subsidy – the difference between the EU domestic price and the world price – will decline. The main influence of the Uruguay Round on the EU cereal sector is likely to be the further restrictions which it will impose upon the payment of export subsidies.

DOMESTIC SUPPORT

There is no uncertainty on the compatibility of the EU's direct compensation subsidies for farmers with the Uruguay Round agreement. The Union persuaded its GATT partners that these subsidies should be regarded as 'production-neutral' and therefore not subject to reduction on the same basis as market support. Had it not done so, then the whole of the return to the EU cereal producer – intervention price plus compensatory subsidy – would have been subject to the 20 per-cent reduction agreed at Geneva in December 1993.

EXPORT SUBSIDISATION

For the Union, the major adjustment to be made under a GATT agreement will be the 21 per-cent reduction in the volume of subsidised exports. Despite

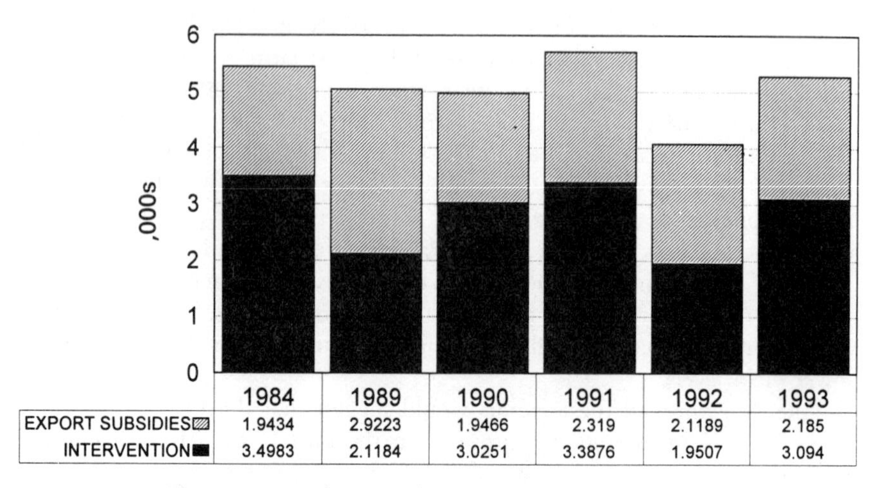

	1984	1989	1990	1991	1992	1993
EXPORT SUBSIDIES	1.9434	2.9223	1.9466	2.319	2.1189	2.185
INTERVENTION	3.4983	2.1184	3.0251	3.3876	1.9507	3.094

Figure 7.5 EC dairy market support budget (MECU)

the 1992 reforms, there could still – when the CAP reforms are fully applied by 1997 – be a significant gap between the (higher) EU market price and the (lower) world price, which would have to be covered by an export subsidy. No matter how small the subsidy might be, the export would be 'subsidised' and therefore subject to reduction.

What would seem likely, therefore, is that, in order to maximise its exports, the Union might develop its own 'export quota' system, with exports beyond the GATT-determined limit taking place without subsidy. There is nothing in the GATT agreement which would prevent the Union from maximising its exports by the operation of such a two-tier export system; indeed, if production exceeds the GATT-set production limit (derived from the export limit) and EU prices do not fall to the world price level, this is what the Union will be forced to do. Alternatively, and more likely, the Union could let the direct support of the cereals market slip to a level close to the world price in the later years of the 'GATT period' and thus eliminate the need for export restitutions.

If the EU maintains a domestic price level higher than the world price it will have to continue to pay export subsidies on at least a part, if not all, of its export deliveries. If all its exports were subsidised, the total quantity would have to be reduced by approximately 16 million tonnes for all grains and for wheat by 7 to 8 million tonnes a year by 2000. This represents a 21 per-cent reduction of the average subsidised export in the 1986–89 period, when the total was considerably less than currently.

Because exports increased substantially during the run-up to the GATT agreement – the period 1991–93 – the actual reduction in wheat exports on

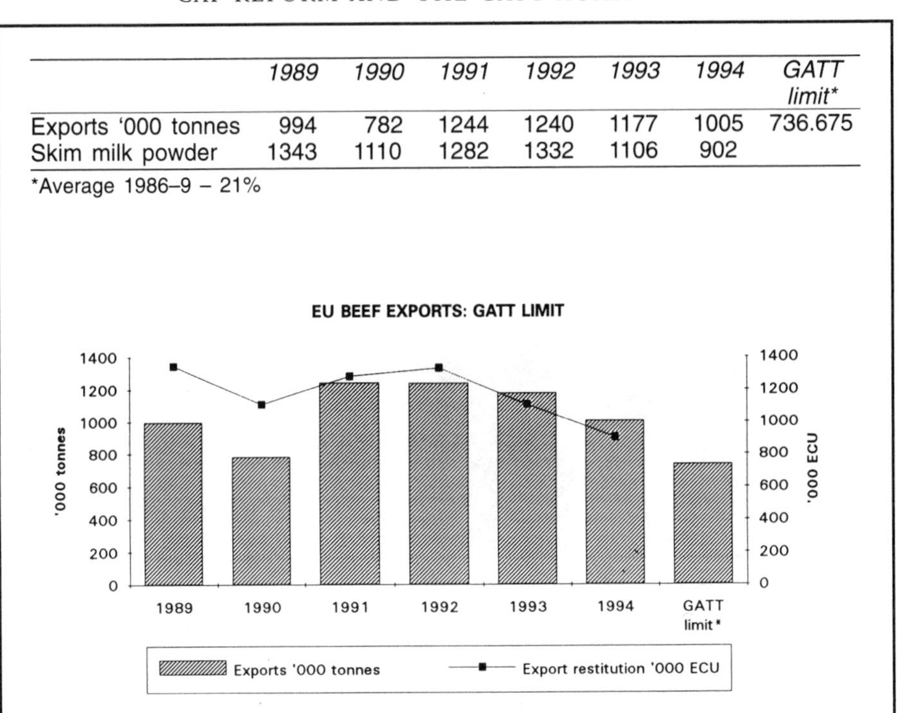

	1989	*1990*	*1991*	*1992*	*1993*	*1994*	*GATT limit**
Exports '000 tonnes	994	782	1244	1240	1177	1005	736.675
Skim milk powder	1343	1110	1282	1332	1106	902	

*Average 1986–9 – 21%

Figure 7.6 EU beef exports: volume and subsidy

the 1992 level will be more than 40 per cent. Under the final agreement reached bilaterally in Brussels and multilaterally in Geneva, the Union will not have to reduce exports as rapidly as originally intended and it will also gain an extra 8 million tonnes of allowable subsidised export in the six-year GATT application period. There are similar concessions for beef, poultrymeat and eggs.

There is too the possibility – some farm management analysts would say the probability – that the EU production cost could fall sufficiently in the later 1990s to allow the support price to be reduced to the world price level without any further increase in compensatory subsidies on the 1996 level. In this situation, the export subsidy is eliminated and therefore too is the GATT limitation on exports.

The hope among EU officials is that by the time the 21 per-cent cut in subsidised exports comes fully into effect, the Union will have been able to get rid of some 10 million tonnes of stocks and that its post-reform cereal price will be down to around 100 ECU/tonne – a point at which, without monetary distortion, the EU could begin to compete at world prices.

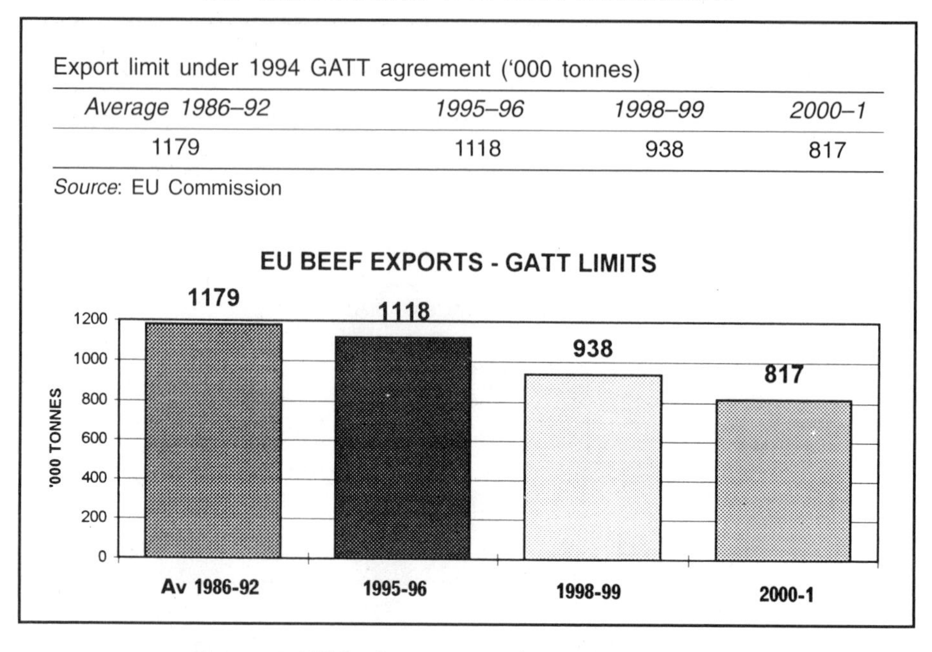

Export limit under 1994 GATT agreement ('000 tonnes)

Average 1986–92	*1995–96*	*1998–99*	*2000–1*
1179	1118	938	817

Source: EU Commission

Figure 7.7 EU beef exports in relation to GATT limits

IMPACT ON THE LIVESTOCK SECTORS

Neither the CAP reform programme nor the GATT agreement will seriously affect the EU dairy industry: the 1984 application of quotas and their extension to 2000 are considered reform enough, and only some small cut in the quota may be needed in the latter part of the 1995–2001 GATT agreement period in order to conform to export limits on cheese and processed products.

The major effect will be on the EU beef sector. While the reform programme has reduced market intervention support, it increased direct production subsidies on breeding (suckler) cows and finished beef cattle; in production terms, the one is expected to cancel out the other.[21] The GATT requirement for a 21 per-cent reduction in exports is the most serious aspect of the changes for the EU beef industry. Because of the very high levels of production and exports in the early 1990s, exports will have to be cut by 31 per cent by 2000. This is likely to require a reduction in production of close to 10 per cent by the end of the century – or alternatively export of a significant proportion of output without subsidies. Either way, this will have a depressing effect on prices and thus can be expected to reduce output around the turn of the century.[22]

The Union's sheep production will not be seriously affected by export limitations, because production is very much less than consumption and exports to third countries are very small. Lamb is one of the few products which

Beef production balance sheet ('000 tonnes)

	Average 1991–	*1995*	*2000*
Production	8560	7958	7777
Imports	540	540	540
Exports	1190	998	817
Consumption	7540	7500	7500
Self-sufficiency – per cent of consumption	113	106	104

Source: UK MAFF

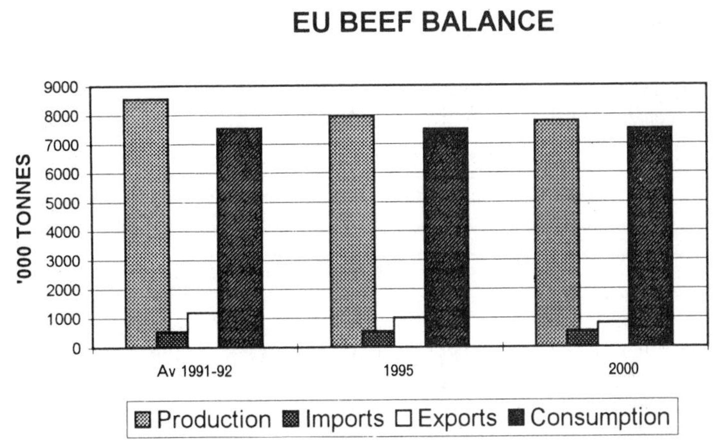

EU BEEF BALANCE

'000 TONNES
9000
8000
7000
6000
5000
4000
3000
2000
1000
0
Av 1991-92
1995
2000
Production Imports Exports Consumption

Figure 7.7 (continued)

the EU imports in large quantities. What will, however, stabilise output – rather than reduce it – are the limitations on the payment of direct subsidies, ewe premiums, to producers[23] under the 1992 reform programme.

REFORM AND THE EU BUDGET

A GATT agreement and CAP reform should both in theory reduce the cost of the agriculture policy. In practice, it is, however, more likely that the reformed CAP will cost more than the policy it is now replacing.

The question which should be addressed therefore is: how far will the GATT agreement moderate the CAP reforms – and more important those elements of the old policy which must be retained – and thus bring about a scaling-down of agricultural support? How far will the GATT limits on EU exports and support contribute to any possible stabilisation of EU agricultural policy expenditure?

The EU Commission's figures[24] quite clearly indicate that the expenditure on the new compensatory subsidies is going to be considerably greater than first estimated when the CAP reforms were originally agreed and that these higher costs will not be compensated for by savings on export subsidies and other support costs – sufficient to keep total spending below the official agricultural support spending limit. The Commission's 1994 estimates indicate that even in 1993 the compensatory payments will have cost as much as 4.3 BECU (billion European Currency Units) and rise to 13 BECU in 1994. Addition of the increments in the arable compensation payments in 1995 and 1996 – plus also inclusion of increases in set-aside payments – will raise the compensation bill to more than 16 BECU.

Against these payments must be set savings on export restitutions and other scaling-down of market support measures. The reduction in cereals institutional and market prices must logically result in a reduction in the level of the export refund, since by cutting the intervention and market price, the Union must inevitably reduce the gap between the EU price and the world

Table 7.1 European Union agricultural budget
(EAGGF guarantee section: 1991–94 plus estimates 1996 and 2000)

	1991	*1992*	*1993*	*1994*	*1996*	*2000*
Cereals, rice/arable	5189	5553	10462	13361	16250	17875
Sugar	1815	1937	2088	2088	3000	3150
Oils, fats/olive oil	5424	5897	2297	1979	2020	2121
Protein crops/dried fodder	959	861	384	364	370	388.5
Fibre plants	522	771	882	742	780	819
Fruit and vegetables	1107	1263	1730	1701	1750	1837.5
Wine	1048	1089	1666	1559	1660	1743
Tobacco	1330	1233	1401	1228	1370	1438.5
Other/other including rice	68	267	402	392	391	410.55
Total crop products	17460	18871	21312	23414	27591	29783.05
Milk products	5637	4008	5184	4050	4600	5000
Beef	4295	4441	4353	4765	4400	3500
Sheep and goatmeat	1790	1751	2267	1568	1850	2000
Pigmeat	252	142	271	194	120	80
Eggs and poultry	169	193	252	177	140	90
Other livestock	0	6	212	155	250	350
Fisheries	26	0	33	36	34	50
Total livestock	12169	10541	12572	10945	11394	11070
Other (inc. food aid and rural dev.)	1280	1555	1798	1505	2000	2200
Total*	30960	31114	35936	35936	40985	43053.05

* Note that grand total does not necessarily sum: this is due to non-commodity, administrative costs and monetary adjust

Sources: 1991–94 European Commission; 1996 and 2000 *Agra Europe*

EC cereal budget projections (million ecu)

	1992	*1993*	*1996*
Exp. refunds	3150	3435	2000
Inter.	2440	2144	1200
Prod. ref.	1028	1113	1112
CRL	– 1126	1126	1126
Comp. subs.	0	1500	6000
Total	5447	9318	11438

Source: 1992 and 1993: EC Commission; 1996: AGRA Europe

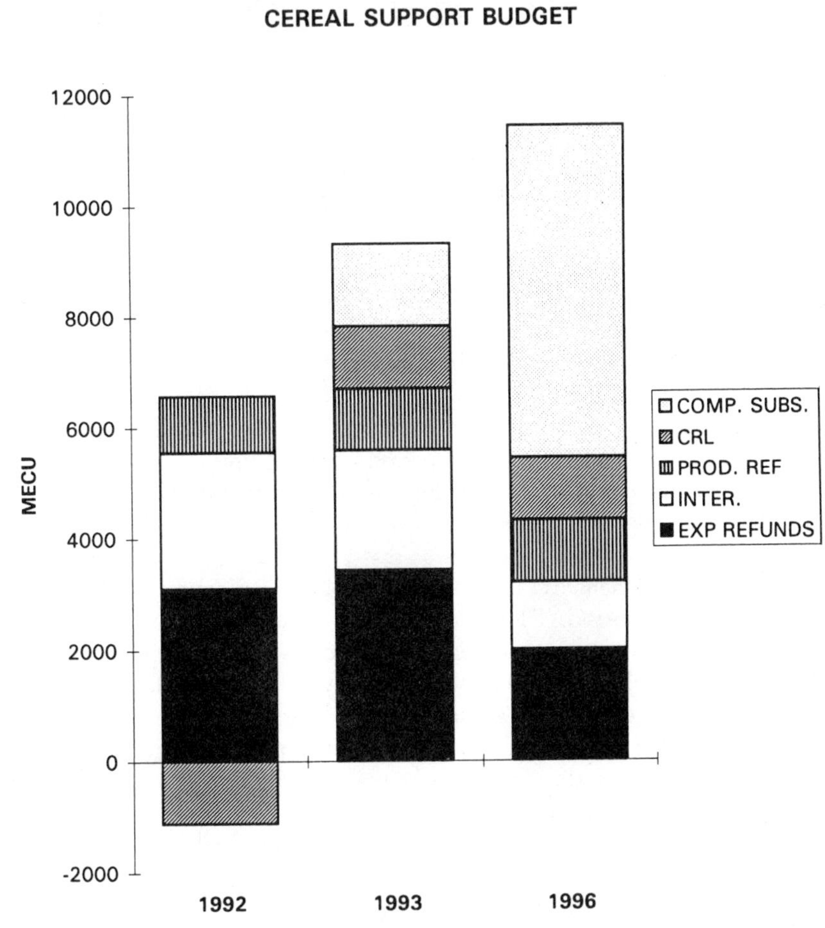

CEREAL SUPPORT BUDGET

MECU
12000
10000
8000
6000
4000
2000
0
-2000
1992
1993
1996
COMP. SUBS.
CRL
PROD. REF
INTER.
EXP REFUNDS

Figure 7.8 EU cereal budget projections (million ECU)

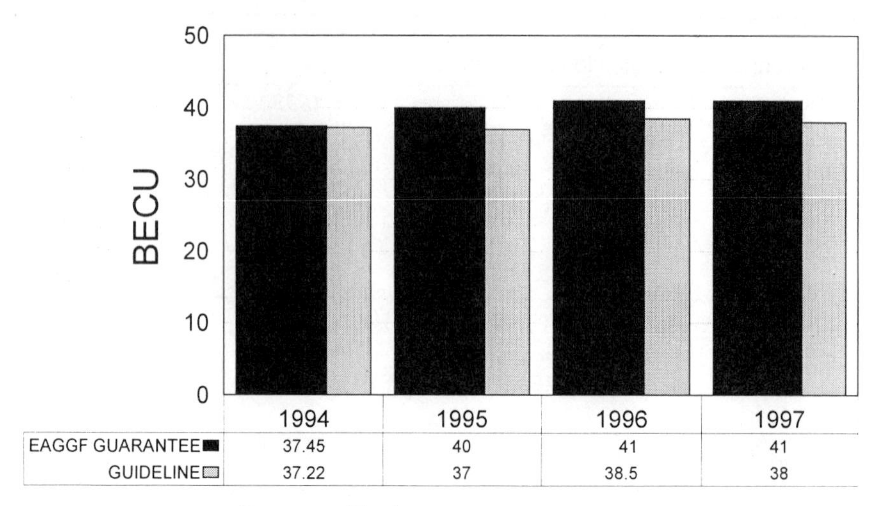

	1994	1995	1996	1997
EAGGF GUARANTEE ■	37.45	40	41	41
GUIDELINE □	37.22	37	38.5	38

Figure 7.9 Total EU agricultural support

Source: EC commission estimates

market price and thus diminish the refund (assuming that the average world price remains unchanged). Crucial to the calculation of the full extent of the saving, however, is the actual volume of exports.

The ability of EU farms to remain viable in this situation of declining prices will depend increasingly on the maintenance of steady rises in productivity. Although the EU's compensatory payments are likely to be maintained certainly until the end of the GATT period – that is, until 1999–2000 – they will not be sufficient to make up the difference between the old pre-1992 prices and the new lower market prices which can be expected during the rest of the decade.

Even though efficient EU farmers are capable of competing at what might be expected to be post-GATT world prices for cereals and grain-fed live-stock products, it is obvious that maintenance of the EU's new compensatory subsidy system is vital to the general profitability of European farming in the future. Maintenance of these subsidies in turn depends upon the extent to which expenditure can be held at a level within expected budgetary limits.

Although it is clear that in the early years of the new systems, the agricultural budget will tend to breach the pre-set limits, it would be reasonable to assume that the Council of Ministers, having forced farmers to make the sacrifice of lower prices under CAP reform, will feel bound to maintain the arrangements and therefore to allow the budget to exceed current limits in the early years of the new CAP/GATT regime.

If, however, the agricultural expenditure level continues to exceed the budgetary limits, then there is a likelihood that there could be a concerted attempt to cut back on the compensatory payments in the later years of a

GATT agreement. This has important implications for the profitability of EU agriculture in the later 1990s and the early years of the next century.

EU production of cereals and beef will tend to decline in the medium term (1993–97) under the influence of, principally, reforms of the CAP introduced in 1992, but also because of the subsequent further impact of the GATT. As a result of GATT restrictions on the subsidising of exports, the volume of the Union's exports of all three commodities will decline during the later 1990s.

The principal effect of the GATT agreement on the world cereal market, initially, will be to raise prices. The reduction in the level of subsidisation of exports by the two major grain exporters, combined with a decrease in the volume of EU deliveries, will raise the world price of wheat and other grains. This rise in the international price will, however, stimulate increased production, not only in the United States, but also in other major exporting countries such as Australia and Argentina. It can also be expected that an initial world price rise will stimulate increased output in the former Soviet Union, central and eastern Europe and in some less developed countries.

Depending on the level of world demand for grain, it can be expected therefore, that the main impact of a GATT agreement will be to increase prices in the medium term (1993–97). The probable rise in world production which this price increase will stimulate is likely to have a longer-term depressing effect on world market prices. The basic underlying reason for this is that world agricultural supply is far more elastic than effective world demand for food.

IMPACT ON FARMERS

There is no doubt that the main effects of the reform on the arable sector has been, as intended, to reduce the area of land devoted to crop production. Its impact on the production of the main crops has undoubtedly, however, been skewed, and it is still unclear what the medium-term effect will be on cereal production trends. The main reason for the reduction in area has, of course, been the requirement for farmers to set land aside if they are to be able to claim compensatory payment (Arable Area Payments – AAPS) to which they are entitled under the 1992 reform programme. The Commission's figures show that the amount of land set aside has increased marginally as the programme has proceeded into its third year, as more farmers have been attracted into the scheme by the increases in the AAPs.

The scheme has, however, had a variable effect on the levels of production of crops within the AAP scheme. Compensatory payments can be claimed only on the land growing cereals, oilseeds and protein (COP) crops. Land devoted to potatoes, sugar beet and other food and non-food crops in the reference year (1992) is excluded. The effect of the differing profitability among COP products has led to an increasing tendency for farmers to maximise the area devoted to the most profitable crops, such as wheat and rapeseed and to the acceleration in the decline in the area of the least profitable crops such as barley. Despite set-aside, the 1995 wheat area is unlikely to be very much less than in 1992 before the inception of the reforms.

More remarkable has been the relatively slight effect of the new policy on individual farmer returns and farm income and the overall increase in farm incomes – the exact reverse of the expected effect. The staged reductions in intervention price for cereals have not been reflected in market prices. A vigorous (subsidised) export policy dissipated some two-thirds of the 1992–93 grain surplus, modest increases in domestic consumption of grain arising principally from high world prices for cereal substitutes (non-grain feed ingredients) and, in some EU countries, substantial green currency devaluations, have meant that there has been little change in the market prices for cereals.

AAPs and set-aside compensation have, however, been paid at the agreed original rates – and in the case of weak currency countries have been increased by green rate devaluations. The result has been that in most countries of the Union the profitability of cereals and oilseed production has increased rather than, as originally expected, been reduced by reform.

Whatever weakening modifications may be introduced by Commission and Council, it cannot be denied that the arable measures have at least stabilised if not actually reduced crop production. The same cannot be said of the livestock sector. The 1992 CAP reforms did nothing to alter significantly the impact of the milk delivery quotas first introduced in 1984; production in 1996 remains substantially in excess of domestic demand. The global level of quotas is unlikely to be sufficiently reduced in the 1990s so as to diminish the Community's dominant role in the international market.

The actual level of surplus dairy output in the EU12 varies according to how the figure is calculated. On the basis of straight comparison between officially recorded deliveries, for purposes of quota administration, compared with domestic consumption, there would appear to be a surplus of around 12 million tonnes of milk equivalent. When, however, various uncounted sources of milk supply, including direct delivery, are included in the calculation, the real level of surplus is likely to be in the region of 15 plus million tonnes. If, in addition, the proportion of domestic consumption of butter, milk powder and other products which are subsidised is added to the total, then it is probable that the Union's real surplus is in the region of 25 to 30 per cent of current 109–110 million tonnes production.

A major problem for the Union is the continuing decline in butter consumption. Butter eating is declining by 1 per cent a year and the decline would, according to the EU Commission, be closer to 2.5 per cent a year, were it not for the substantial subsidies on consumption. These subsidies – amounting to over 350 million ECU a year – affect 30 per cent of the butter consumed in the Union.

Originally, the Commission's 1992 reform package recommended cutting the quota by 3 per cent – some 3.3 million tonnes and 22 per cent of the then 15-million-tonne milk surplus. The recommended reforms also included a 15 per-cent cut in the butter price and a 5 per-cent reduction in support levels for skimmed milk powder (SMP). The Council of Ministers managed to water down this very modest adjustment to an undertaking to 'consider reducing the quota by 2 per cent over two years'. The butter price was cut by only 3 per cent and there was no change in the SMP price. The Council and the Commission subsequently managed effectively to increase the milk quota above the level at which it should have been had the reforms been applied.

Decisions in 1994 and 1995 to increase the Italian, Spanish and Greek quotas will have the effect of reinstating all of the modest reductions in quota which should have been applied during the reform period. After the 1995 changes, the total global delivery quota will still be 109.05 million tonnes – only 0.56 million tonnes short of the 1992–93 quota level. There will, however, have been significant reductions in the national quotas for Germany, France, the Netherlands and the United Kingdom.

In the beef sector, cyclical and technical factors have played a more important part in reducing production than any measures taken by Commission or Council. The Union's beef production is likely to continue the trend of the early 1990s: a modest reduction in output in response to falling real prices. What is needed, however, is a significant reduction in output right through the rest of the decade if the EU is to avoid either breaching the GATT limit or again building chronic and increasingly unsaleable domestic stocks. The problem is not only one of too high production, but also of declining beef consumption. While if the early 1990s trend of production continues output will decline by little more than 3 per cent by 2000, consumption will also decline if present trends continue, by close to 4 per cent. The incipient imbalance in the market is thus likely to remain unchanged.

Apart from the failure to reduce the budgetary cost of supporting the beef sector, which will also remain undiminished at its current 5 billion ECU level – since the cost of direct subsidies will more than counterbalance any modest savings on export or other market support subsidies – these policy changes will do nothing to adjust output to the increasingly serious GATT limit on subsidised exports. Under the GATT accord, the Union must reduce its beef exports by 60,000 tonnes a year to achieve a maximum export of 817,000 tonnes by 2000. This figure is close to 400,000 tonnes less than the estimated 1.2-million-tonne export of 1994.

What the figures for all the main commodity sectors would suggest is that, as the 1992 refrom programme approached completion in 1996, the Union was operating on a knife edge, where climatic variation, world market price levels or internal technical changes could tip the balance of production beyond the limits of budgetary cost and permissible exports.

Gross value added by sector

Cereals EU12	1985–89	GVA per hectare ECU 1991	1992	1993	1994
Without subsidies	502	536	499	446	333
With subsidies	502	536	499	621	569

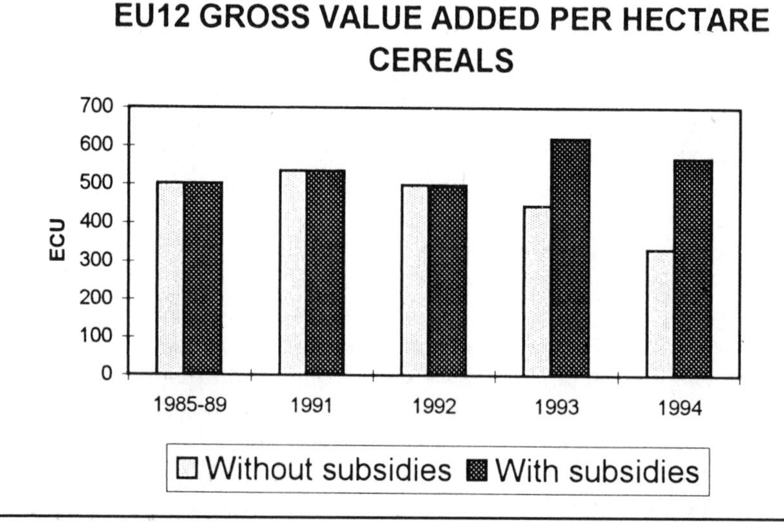

Figure 7.10 Impact of reform on cereal producers' returns

The impact of the new policy on farm incomes has been surprising: farmers' profits have in general been increased rather than been stabilised or reduced – as had been expected. But what has become clear is that the new system of supporting farmers has created a system of agricultural support where direct subsidies from the European Union represent the difference between profit and loss for the majority of farmers. A European Commission analysis on EU farm income trends in the period 1985–94[25] shows quite clearly that incomes are only being sustained in real terms by the compensatory subsidies for arable crops and grazing livestock payable under the 1992 reform programme.

The SPEL system[26] analysis showed that, aggregated for the Union the proportion of income represented by the new subsidies had reached more than 25 per cent of the net value added of the agriculture sector by 1994. Since that was the first full year of the reform programme, it can be expected that the proportion will rise significantly – to probably 35 per cent plus – by the last year of reform in 1996. This assumes that the market prices will by then fully reflect the 30–34 per-cent reduction in intervention price levels

Gross value added by sector

	GVA per hectare ECU				
Oil seeds EU12	*1985–89*	*1991*	*1992*	*1993*	*1994*
Without subsidies	681	472	3	62	137
With subsidies	681	472	550	588	662

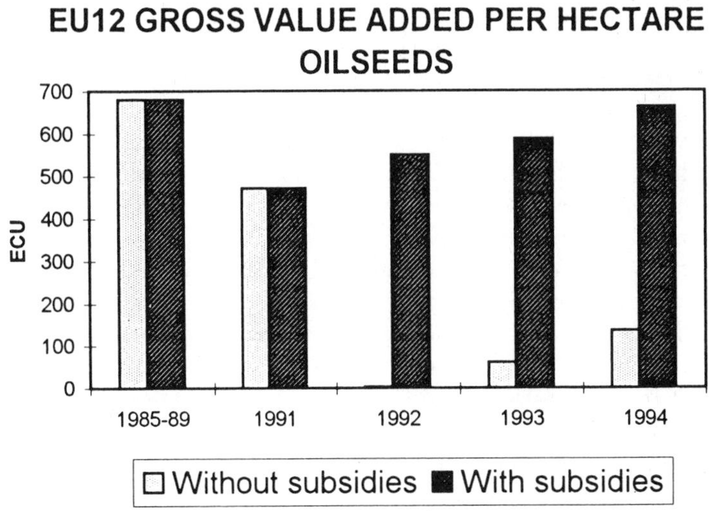

EU12 GROSS VALUE ADDED PER HECTARE OILSEEDS

ECU
700
600
500
400
300
200
100
0
1985-89
1991
1992
1993
1994
Without subsidies
With subsidies

Figure 7.11 Impact of reform on oilseed producers' returns

agreed for the end of the reform period. If market prices continue to remain significantly above the intervention level, then total returns will increase on the 1994 level, since the increased subsidies will continue to over-compensate for the official support price reduction.

Also significantly, the figures show that farm incomes were approximately 6 per cent higher in the 1990–94 period than in the previous five years – largely due to the income boost from subsidies in the last year of the period.

In the cereal sector, for example, the average gross value added (at market prices) of output declined from approximately 500 ECU per hectare in 1992 to 330 ECU in 1994. When however compensatory subsidies are added, the GVA rises to 570 ECU/ha – an 8 per-cent increase in real terms on the 1992 level. Significantly, there was a large reduction in average production costs for cereals of 11.4 per cent in this period. The importance of the subsidy is even greater in the oilseeds sector, where the market price for EU production is much more closely related to world prices: while the market value of the output fell by a massive 73 per cent between 1991 and 1994, GVA actually increased by 29 per cent in real terms.

The main livestock sector affected by the 1992 reforms, beef, received nothing but net gain from the new system. Since market prices for beef in the 1991–94 period were largely unchanged, payment of the new beef special premium and suckler cow premiums resulted in a 24 per-cent increase in real gross value added per head of beef animal produced between 1992 and 1994.

What the SPEL analysis also shows – less obviously, however – is that the rise in incomes was also caused by additional factors which were probably largely due to the indirect impact of reform on producer attitudes – most notably the reduction in input use, the fall in the price of inputs, and in some countries of the Union, a reduction in the agricultural labour force. The expectation of reform and the 'shock' of the reform itself led to a 13 per-cent reduction in the use of plant protection products and a 12 per-cent reduction in average fertiliser use. For arable crops the reduction was even more startling: a reduction of 20 per cent for plant protection chemicals and 15 per cent for fertilisers. The reduction in demand for these inputs also had the predictable effect of reducing their price.

Clearly, the reform has the most significant effect on arable cropping and grazing livestock; only indirectly does it affect grain-fed livestock, dairy production (so far unchanged by the 1992 programme) and fruit and vegetables. Those sectors which are affected do, however, represent a major proportion of the total agricultural aggregate income.

These figures show not only the importance of the new subsidy system to the maintenance of agricultural incomes, but also the extent to which average incomes are also being sustained by productivity increases arising principally from the shedding of agriculture labour and the more efficient use of variable inputs. The SPEL figures would indicate that where this has not taken place – most notably in France – incomes are still declining despite the buttress of the new subsidy system.

The danger for the agriculture industry in the longer term is of course that the apparent guaranteed income provided by the reform programme will encourage the erosion of the benefit of the subsidies to producers through rising land prices and increased prices for inputs. There are already clear signs that this is happening: the price of prime arable land in France and the United Kingdom has risen more than 30 per cent since the early 1990s, due principally to the guarantee of subsidy income which the holding of such land brings, and there are already signs that fertiliser and crop-protection chemical manufacturers are raising their prices as production has adjusted to the new, lower demand levels from the farming industry.

This vulnerability of the industry to cost increases was illustrated in a report on CAP reform effects on the UK arable sector by economists at the Department of Land Economy at Cambridge University.[27] Fertiliser and crop chemical costs are now on a rising trend, as are land prices and rents, they point out. 'These factors will increase the unit cost of cereal production and reverse the trend of cost containment, which has been apparent since the mid-1980s.'

The important political conclusion to be drawn from these figures is that the whole structure of new subsidies erected by the MacSharry reforms, instead of being merely transitional mechanisms to adjust EU agriculture to post-GATT price levels will become more rather than less necessary to the maintenance of the viability of the industry. It will be as difficult to remove the compensatory system in the years after 2000 as it was to scale down the intervention system in the 1980s.

There is little sign either that the new EU policy or the GATT have cut the burden of agricultural policy to taxpayers or consumers. The 1994 OECD report on the operation of agriculture policies[28] shows no reduction in cost – but rather an increase. The OECD indicates that the level of support – most observers would say over-support – to farmers has if anything increased since 1993, while food prices have not fallen in line with supposed reforms.

The Organisation says that a total of $175 billion was spent on subsidies to farmers in OECD countries in 1994 – an increase of $8.5 billion compared with 1993. But because the overall value of farm production also increased, the level of public subsidy expressed as a proportion of total output (the Producer Subsidy Equivalent, or PSE) remained unchanged at 43 per cent. The Consumer Subsidy Equivalent, which measures the cost to consumers of protection and support, declined by around $2.6 billion to some $123 billion, a CSE rate of 33 per cent.

In the European Community, the OECD report confirms that the switch away from state-financed market manipulation to direct subsidies has meant that the overall level of government handouts to farmers has remained largely unchanged. Only the form of subsidy payments has changed. Direct subsidies in the EC have increased by 25 per cent between 1992 and 1994.

> 'The preliminary producer subsidy estimates indicate that there has been a further slight increase in assistance to the EC cereal sector in 1994, following the sharp increase which occurred in 1993. This is mainly because the policy is designed to provide full compensation for support price reductions.' The report supports the European Commission view that these so-called compensatory subsidies over-compensate for market price reduction. In general the reform support price reductions have not been fully reflected in producer prices. With respect to beef, the rise in direct payments has also more than offset the fall in the producer price.[29]

The Community was responsible for close to half of the total $80.48 billion OECD transfer of resources to agricultural producers in 1994. The EC's percentage PSE also rose by one percentage point compared with 1993, to 50 per cent. The United States' aggregate PSE, on the other hand, fell to a low of 21 per cent, from 23 per cent in 1993 and 30 per cent in 1986–88. The degree of subsidisation per tonne of output in the EU was higher in 1994 than in the year before, although the volume of production was lower. Market price support fell on a per-tonne basis, but the OECD noted a rise of more than 20 per cent in unit direct payments, as a result of the introduction of CAP reform compensatory payments. Subsidies paid direct to farmers accounted for 20 per cent of total assistance to the sector, but market price support still accounts for around 65 per cent.

The May 1992 reforms of the CAP and the 1994 GATT agreement on international food trade reform will not give massive gains in consumer benefit. Independent experts say that the most that can be expected is a return to the levels of consumer loss – resulting from paying too high a price for food – that ruled in the early 1980s,[30] before the current phase of over-protection of farmers really took off.

It has been estimated that the tax on the consumer is as high as 370 per cent in Japan, 120 per cent in the EFTA countries, 75 per cent in the EU and 5 per cent in the United States. The gains from liberalisation of world agricultural trade would appear to be substantial. According to the conventional view, not only would the 'protection tax' at current prices be eliminated, but the prices to be gained by farmers from the world market would also substantially increase.

The total costs to the consumer of the CAP is estimated by the OECD for 1994 at 50.7 billion ECU – the equivalent of 42 per cent of the total cost of food in the European Union. Although the cost is not as high in percentage terms as in Japan or some of the EFTA countries, the Union holds the unfortunate record for paying out the largest amount both from taxpayers and consumers to its farmers. The total producer subsidy equivalent to the EU farm industry is close to US$100 billion – nearly twice the amount paid out by consumers and taxpayers to American farmers – despite their much larger industry.

Calculations on the basis of the OECD's figures for the Consumer Subsidy Equivalent (CSE) or consumer welfare loss, and adjusted on the assumption of the GATT agreement involving a 36 per-cent reduction in tariffs in the period 1994–99, would suggest that the reduction in welfare loss would be substantial. It is, however, a sobering thought that even at the end of the current decade and assuming a 'second stage' to the Uruguay Round from 1998–99 for a further five years, the CSE in the year 2000 for the major products would only be back to where it was in the early 1980s.

PART III

EUROPEAN AGRICULTURE AND THE FUTURE

8

EUROPEAN FARMING AND THE ENVIRONMENT

> If we are to curb the grinding poverty that afflicts over a billion people today, and cater for a population increase from today's 5.8 billion to the projected 8.5 to 10 billion in the second half of the next century, we shall have to about double world food production. And we shall have to do so sustainably – without losing more Indias. This is likely to mean great increases in production from the more fertile lands suited to intensive cultivation rather than more encroachment on the land that retains natural vegetation.
>
> Martin Holdgate (former Chief Scientist, UK Department of the Environment)[1]

Holland has a major new and rapidly expanding export product: manure. Each year larger quantities of pig, poultry and cow manure are dried, bagged and shipped out of the Netherlands to be sold as fertiliser in garden centres all over Europe. The process is expensive, and in the normal sense of the word, 'unprofitable'. The people of Holland have, however, decided that the environmental costs of continuing to poison their drinking water with animal effluents justify the hefty government subsidy which is now being paid to the companies processing the animal waste surplus into the latest fix for roses and begonias. The Dutch Government has decreed that the manure disposal plants must have a 200-million-tonne annual capacity by 2000.[2]

Most of the other northern countries of the Union face the same problem on a similar or smaller, but none the less serious, scale: too many people and too many animals all mixed up together on too little land. So far, none of them has yet taken the sweeping steps already taken by the Hague. But they will, as tighter EU legislation on water contamination is imposed. The Dutch example illustrates in a broader sense the major challenge which European agriculture faces in the 1990s: how to maintain production and profits with less pollution.

While farmers seek the latest scientific innovations to reduce their costs and maintain their incomes against a trend of falling real prices for their produce – as more reforms are forced on the CAP, they will be constrained from maximising production by limits on the use of fertilisers and pesticides and habitat-damaging techniques. There will probably also be increasingly tight limits on the use of many of the new biotechnological techniques which might have allowed them to cut costs.

Agriculture is in the forefront of public concern about the state of man's physical surroundings. In Europe and in the European Union, the farmer is now seen as a major polluter and destroyer of what urban man perceives as the 'natural' environment. Along with the air- and water-polluting chemical industry, the agricultural industry is perceived by the urban majority as over-using fertilisers and crop protection chemicals – and thus polluting water supplies and the very food that we eat.

Europe's environmental problem to a great extent arises from the concentration of population on relatively small areas of land. Major damage was done to large parts of western Europe's landscape and environment by industrialisation in the nineteenth and twentieth centuries. Agriculture, on the other hand, has only recently 'caught up' in the environmental damage stakes; its effect is therefore all the more spectacular for being more recent.

Since the early 1950s, European agriculture has become mechanised, more intensive, thus employing more fertilisers and pesticides, and in the process changing the landscape and too often destroying not only wild animals and plants but also the habitats in which naturally occurring fauna and flora once throve.

In the main, this change is the response to demand. As more people have demanded more and better food, so Europe's agriculture industry has responded by producing more. The pollution concentration in the north-western countries of Europe is among the highest in the developed world. This large human population density has brought with it an increasing density of intensive livestock – pigs, poultry and dairy cows – but also led to the intensification of arable agriculture. Alongside man's increasing effluvia of sewage and chemical wastes has grown the large outflow of nitrates, phosphates and other possible pollutants from intensive livestock production. Effluent from intensive livestock production is now taking over from the chemical industry as the major source of river and groundwater pollution.

Reduction and prevention of such pollution is now a major concern in all the densely populated areas of the European Union: the whole of the Netherlands, the low-lying areas of Belgium and France, most of southern England, large areas of Germany and the northern plains of Italy.

The main environmental risks or hazards arising from modern agriculture in developed countries and Europe in particular are: (1) pollution of water supplies by nitrates and to a lesser extent phosphates; (2) wider damage to the environment arising from the over-use of fertilisers and run-off from intensive livestock production; (3) the damage to wildlife and plants resulting from the over-use or misuse of pesticides; and (4) damage to human health resulting from the misuse of pesticides.

The damage to the environment from the over-use of fertilisers arises primarily from the tendency of farmers to over-dose with plant nutrients in order to maximise yield. Measuring the actual impact of the activities of agriculture on this undoubted soil contamination process is, however,

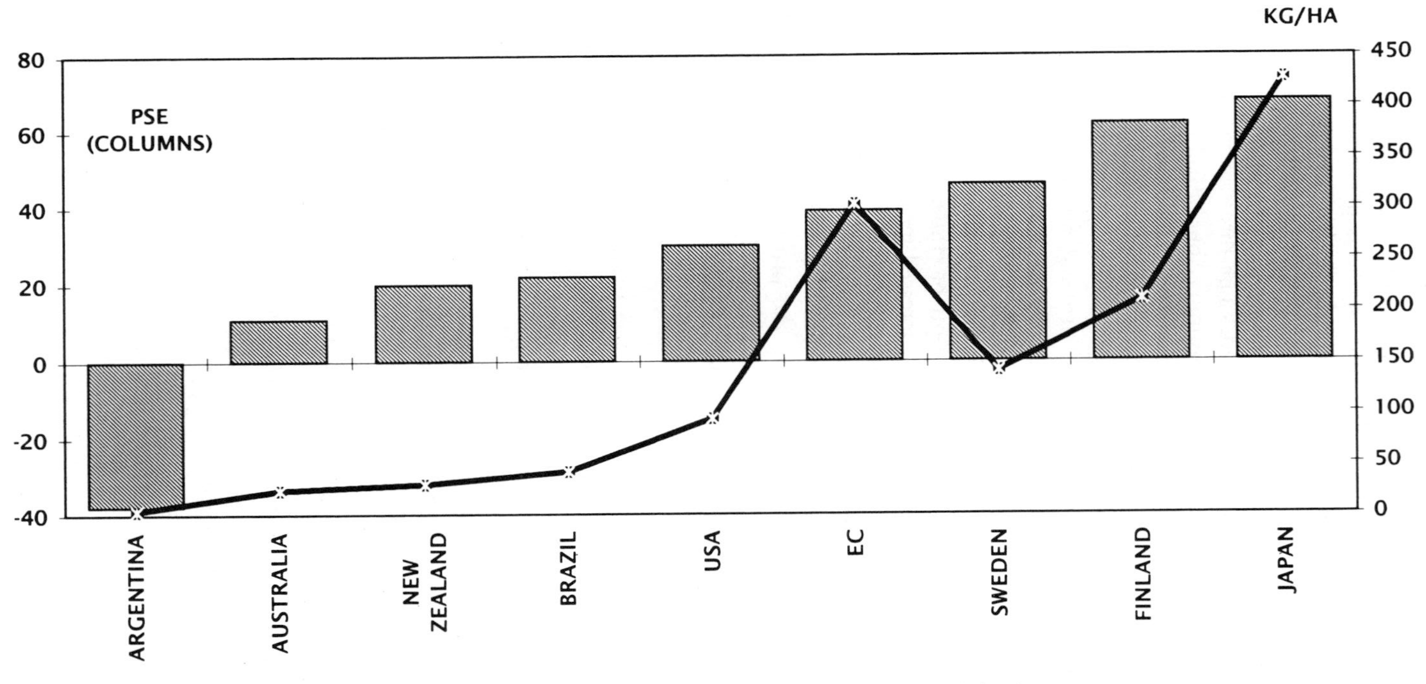

Figure 8.1 Relationship between farm support and fertiliser use

Source: US Department of Agriculture

extremely difficult because a considerable proportion of the nitrogen run-off from farmland results not from the 'artificial' fertilisers or animal manure applied by the farmer, but from the natural reserves of nitrogen which are already present in the soil.[3] It is none the less likely that agriculture is the major cause of environmental contamination by nitrates and phosphates. It has been argued that over 50 per cent of the nutrients applied by farmers to crops in Europe and other developed regions of the world are wasted and become an environmental hazard.[4] The threat to the environment from fertiliser arises not from direct poisoning, but usually from the insidious change in the state of vital features of the landscape such as rivers, lakes and coastal waters, usually arising from eutrophication resulting from nitrate run-off. Compared with the change in plant populations in pastures and meadows stimulated by the use of artificial fertilisers (which can be relatively easily put right), the longer-term damage to waterways is far more important.

The two claimed human health hazards arising from the over-use of fertiliser – and the uncontrolled run-off of effluent from intensive livestock production – are the possibilities of increased incidence of gastric cancer and 'blue baby' syndrome – methaemoglobinaemia.

The problem of livestock effluent run-off arises from the sheer concentration and volume of the problem: the output of large, intensive livestock units is far greater than can be absorbed by the plants grown on the surrounding farmland. It has been estimated[5] for example, that a modern (relatively small) farm of 40 hectares with a dairy herd of fifty cows and a fifty-sow pig herd has a potential 'pollution load equivalent to that of a village of 1000 inhabitants'.

Throughout the period of post-war agricultural intensification the problem of intensive livestock effluent has grown apace. In the low countries of western Europe, in particular, the effluent output of the livestock industry has long since exceeded the capacity of arable and grassland to absorb it. In almost all areas of dense human population and therefore heavy intensive livestock population there is an animal waste 'surplus' which is too often now adding to the already high levels of nitrogen and phosphates in watercourses, lakes, seas and groundwater.

Table 8.1 Cattle and pigs per UAA for selected EC countries, 1989

	Belgium	*Netherlands*	*Denmark*	*Germany*	*France*	*Italy*	*UK*
			Head per UAA[1]				
Cattle	2.29	2.36	0.79	1.23	0.7	0.51	0.65
Pigs	4.75	6.80	3.27	1.90	.45	.54	.40

Source: USDA

[1] Utilised agricultural area

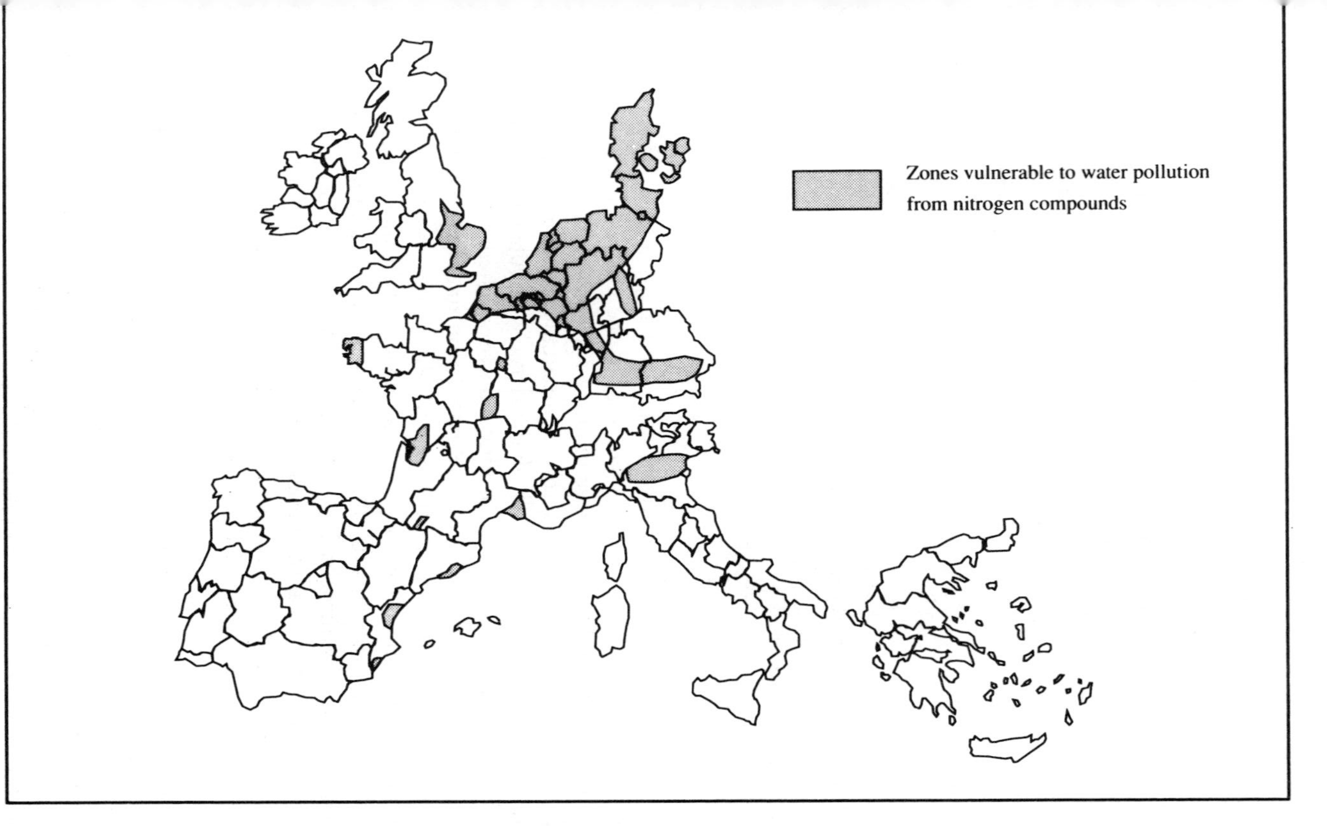

Figure 8.2 Map showing nitrate-sensitive areas of EU

Source: Agra-Europe 52/88, 27 December 1988

Table 8.2 Nitrogen fertiliser applied and uptake by crops, EC10, 1986

	Nitrogen uptake, all sources							
Nitrogen fertiliser								
Country	*Wheat*	*Coarse grains*	*Straw*[1]	*Rice*	*Forage*	*Total*	*Applied*	*Residual*
				1000 metric tonnes				
Belgium[2]	25	16	10	0	160	211	199	–12
Denmark	41	87	32	0	127	287	381	94
Germany	195	230	106	0	783	1314	1578	264
Greece	49	43	23	1	287	403	432	29
France	505	346	212	1	1342	2406	2568	162
Ireland	8	23	8	0	368	407	343	–64
Italy	18	131	76	14	633	1027	1011	–16
Netherlands	0	5	6	0	256	284	504	220
UK	263	159	105	0	995	1521	1671	150
Total	1104	1040	578	16	4951	7860	8687	827

Sources: O. J. Leuck, USDA, Washington DC, 1993

1 It is not clear if the coefficients for straw nets out the nitrogen that is returned to the ground because of decomposition.
2 Also contains a small amount for Luxemburg.

The pollution of groundwater, rivers, lakes and estuaries by the run-off from intensive livestock and arable farming is the major environmental problem for which the EU agriculture industry is responsible. Intensive livestock production is, however, a far more important source of this type of pollution than cropping. Almost all of the densely populated areas of the Union also have intensive livestock farming industries. There is thus a conflict between the food supply needs of that population and its need for clean water free of contamination with nitrates and phosphates. Livestock effluent disposal rates on the most densely populated areas of the Union are between three and four times what are officially designated as the 'maximum acceptable concentration' (MAC) nitrogen levels. A glance at a recently compiled map (Figure 8.2) shows that areas of animal effluent surplus equate with the areas of densest human population. At its worst in these areas the nitrogen surplus is estimated to be ten times the quantity that it is possible for plants to absorb.[6]

Contamination of water is also being increased by heavier fertiliser applications to arable land. During the last forty years the usage of nitrogen on crop land and pasture has increased by over 400 per cent in the EU10. The current average application rate of 100 Kg/ha suggests that the application rates in the most intensively farmed areas are likely to be as much as four times this figure.[7]

What this increased usage of fertiliser and greater disposal of animal wastes add up to is an increasing imbalance of nitrates in the soil. For example, according to one authority,[8] while the 'annual nitrogen surplus' remaining in the soil in Germany was no more than 10 kg/ha of farmed land twenty years ago, today it is more than 100 kg N/ha. Germany is less intensively farmed than the Low Countries or parts of England and France; it can therefore be deduced that the nitrogen imbalance in these areas is likely to be even greater. Increasing 'soil surpluses' of phosphate and potash are also being noted. Largely because of this agricultural effluent, it is estimated that 5–6 per cent of the European population is now being supplied with drinking water which contains more nitrate than the permitted EU maximum of 50 mg NO_3/litre, and 25 per cent of the population is using water with a level greater than the optimum of 25 mg/l.

Water supply administrators are finding the levels of nitrate contamination rising and thus more difficult to control. In the German Land of Baden-Wurttemberg the authorities report that the percentage of water exceeding the EU maximum has risen from 17 to 21.8 per cent during the first half of the 1980s. Every year the number of catchment areas capable of delivering water with safe nitrate levels decreases. But the problem is not only the limitation of contamination, but also of dealing with the concentrations already present in the soil. Scientists point out that there is a ten- to twenty-year lag before nitrate ends up in the groundwater after being washed from the topsoil.[9]

Farming has officially been identified as the major cause of contamination of watercourses by nitrates and to a lesser extent phosphates. Almost all of the densely populated areas of the Union have intensive farming industries. European Commission studies show quite clearly that the livestock effluent disposal rates on the most densely populated areas of the Union – in the Netherlands and Belgium particularly – are between three and four times what scientists regard as MAC nitrogen levels.[10]

Areas of what the Commission politely calls 'animal effluent surplus' equate with the areas of densest human population. At its worst in these areas, the nitrogen surplus is estimated to be ten times the quantity that it is possible for plants to absorb. From this it follows that the surplus seeps into the groundwater and thence into the human drinking-water supply.

The EU's Council of Ministers agreed in June 1991 a Directive designed to limit the nitrate levels in water supplies, which could seriously limit output and cut farmers' incomes in some of the most intensive livestock and arable farming areas of the Union by more than 50 per cent. Though it could reduce nitrate contamination of water and have other beneficial effects on Europe's rural environment, it could have a devastating effect on farming in certain areas.

To conform to the planned EU limits, the rate of nitrogen application to arable land would have to be cut to 75 kg N/ha – about half the current average rate of application. This would reduce farm incomes by an average 57 per cent, it has been expertly estimated.[11] Intensive livestock farmers –

Table 8.3 Nitrogen fertilizer produced from livestock manure[1,2]

Country	*Dairy*	*Beef*	*Pigs*	*Layers*	*Broilers*	*Sheep*	*Total*
				1000 metric tonnes			
Belgium[3]	65	137	122	5	41	10	381
Denmark	58	109	224	2	40	1	435
Germany	349	651	549	25	102	42	1717
Greece	14	36	33	8	32	332	455
France	416	1043	275	33	299	327	2393
Ireland	98	272	30	2	15	120	536
Italy	197	380	154	23	138	265	1157
Netherlands	149	176	249	19	143	16	752
United Kingdom	208	604	217	25	254	511	1819
Total	1555	3408	1851	142	1064	1625	9645

Source: O. J. Leuck, USDA, Washington, DC, 1993

1 Numbers may not sum to totals because of rounding. Beginning inventories are used for cattle; number slaughtered are used for pigs and sheep, with 7% and 50% being added to account for the breeding herd; and the number of eggs hatched for chick placement for eggs and meat are used for layers and broilers.

2 Also contains a small amount for Luxemburg.

the broiler chicken, battery hen, cows on concrete and pig factory men – would have to cut back even more.

The Nitrates Directive,[12] is designed to restrict the leaching of nitrates in what are classified as 'vulnerable' zones where there is most conflict between arable and livestock production and the need for nitrate-free human water supplies. The objective is to keep the amount of nitrate in water supplies to below the maximum allowable concentration of 50 parts per million laid down in the Union's original Drinking Water Directive of 1980. The intention of the Directive is that the level of 'residual nitrogen' – the amount of nitrate remaining in the soil and soil water after normal absorption by crops – should be reduced by 4 to 5 per cent.

The Directive is to be brought into operation over the eight-year period to 1999. Member states have designated their vulnerable zones where measures will have to be taken to reduce the concentration of manure disposal and nitrate surplus from arable farming operations. They are (in 1996) in the second stage of application of the Directive: drawing up specific policies to keep nitrate contamination below the MAC.

In general, this involves the operation of codes of 'good practice' by farmers in the designated areas. In some areas of the Union where the nitrate 'surplus' is likely to be at its highest, however – such as in most of the Netherlands and in large parts of Belgium, Denmark, northern Germany and the United Kingdom – special measures are having to be introduced to remove the

Table 8.4 Nitrogen use, uptake and residual, EC10, 1986

Country	*Use* Livestock	Inorganic	Total	*Uptake*	*Residual* Total	Kg per ha	Share of total
	1000 metric tons					*Kg*	*(%)*
Belgium[1]	382	199	580	211	369	240	64
Denmark	434	381	816	287	529	187	65
Germany	1717	1578	3295	1314	1981	165	60
Greece	455	432	887	403	484	84	55
France	2393	2568	4961	2406	2555	81	52
Ireland	536	343	879	407	473	83	54
Italy	1157	1011	2167	1027	1140	65	53
Netherlands	752	504	1255	284	972	480	77
UK	1819	1671	3490	1521	1969	106	56
Total	9645	8688	18333	7860	10473	1491	57

Source: O. J. Leuck

[1] Also contains a small amount for Luxemburg.

animal effluent 'surplus' from the region. These measures will be implemented in the period 1996–99. These areas have the greatest concentration of livestock: the Netherlands, for example, has 6.80 pigs per unit of utilised agricultural area (UAA), Belgium 4.75 and Denmark 3.27.[13]

The calculation of the nitrate surplus above the MAC which has to be eliminated must take account of the amount of animal effluent which is being disposed of on the land, the amount distributed as chemical fertiliser and the amount of nitrogen already in the soil.[14] Crudely calculated, this amounts to a situation where the maximum annual residual (MAR) after taking into account the utilisation by crops, must not exceed 170 kg of nitrogen per hectare. Local authorities are having considerable difficulty in

Table 8.5 Residual nitrogen and reductions in livestock and fertiliser use to achieve the maximum annual residual (MAR), 1986

Country	*Nitrogen residual* Above MAR	Reduction to MAR	*Reductions only in livestock*	*Reductions in both livestock numbers and fertiliser* Livestock	Fertiliser
	1,000 metric tons		*(%)*		
Belgium	107.0	29	28	28	0
Denmark	47.6	9	11	9	2.2
Netherlands	632.0	65	84	65	28.0

Source: O. J. Leuck

Table 8.6 EC10 livestock numbers: base, nitrate scenario, 1991

Item	*Dairy*	*Beef*	*Pigs*	*Layers*	*Broilers*	*Sheep*
	1000 head				*Million head*	
Belgium:						
1986 base	1021	2142	9363	525	10.9	85.8
Nitrate limit	724	1519	6638	372	7.7	60.8
1991	893	2483	9073	560	10.9	110.8
Denmark:						
1986 base	913	1710	17228	60	4.0	82.8
Nitrate limit	827	1550	15617	54	3.6	75.1
1991	769	1472	17565	107	4.1	99.5
Netherlands:						
1986 base	2333	2743	19158	788	40.6	297.3
Nitrate limit	826	971	6782	279	14.4	105.2
1991	1917	2913	21337	1050	41.4	319.1
Other EC:						
1986 base	20036	46652	96661	79855	240.7	1750.0
Nitrate limit	20036	46652	96661	79855	240.7	1750.0
1991	17440	45444	96500	88286	225.7	2197.0
Total:						
1986 base	24303	53247	142410	81228	296.2	2215.9
Nitrate limit	22415	50696	125720	80489	266.4	1991.2
1991	21019	52312	144475	90003	282.1	2726.4

Source: O. J. Leuck

the interpretation of this legislation not only because of the many variables which affect the nitrate level in soil and water, but also because they are continuously changing with soil, atmospheric and plant growth conditions.

The extent of the changes that might have to be made by the EU agriculture industry to conform with the Nitrates Directive is thus extremely difficult to calculate. The only detailed study of this problem was carried out in the mid-1980s,[15] but levels of animal effluent disposal and fertiliser use are unlikely to have altered by any large amount since then, although the average level of fertiliser use in the EU10 has decreased since 1988.[16]

This study indicated quite clearly that the nitrogen residue left behind by arable farming activities is relatively small: of the 8.7 million tonnes of nitrogen applied to the major crops in the EU10 in 1986, approximately 10 per cent, 828,000 tonnes, would have been left in the soil at the end of the crop-growing process. This indicates very strongly that the major problem lies with intensive dairy, pig and poultry units. The problem is that, though the nitrate surplus from crops may be small, almost all of that coming from

Table 8.7 Reduction in livestock to achieve the MAR, 1986 (percentages)

Item	*Dairy*	*Beef*	*Pigs*	*Layers*	*Broilers*	*Sheep*
Belgium	4.2	4.0	6.6	3.7	3.9	0.60
Denmark	3.8	3.2	12.2	1.4	3.7	.07
Netherlands	9.6	5.2	13.5	13.7	13.4	.97
Total	17.6	12.4	32.3	18.8	21.0	1.64
Change in EC9 livestock[1]	7.8	4.8	11.7	10.1	10.1	.91

Source: O.J. Leuck

[1] Total percentages may not agree because of rounding.

livestock production must be surplus and a threat to the MAC limit. This has been calculated to be more than the whole of nitrate applied to crops – some 9.6 million tonnes. In crude terms, this means that the EU10 has an approximate 57 per-cent 'nitrogen surplus'; the national figures vary from 52 per cent in France to 77 per cent in the Netherlands.[17] The problem is greater in countries such as the Netherlands, Belgium and Denmark, where the nitrogen surplus from cropping is already high and where there are heavy concentrations of intensive livestock production.

Since the areas of greatest nitrate surplus from agricultural activity coincide with what are being designated as vulnerable zones, it is clear that significant changes will have to be made in the farming pattern or in the manure disposal policies – or both – pursued in these areas. The onus is going to be largely on the livestock sectors to make the necessary changes. The Directive allows the nitrate surplus to be reduced by cutting either livestock effluent distribution or crop dressing with fertilisers or a combination of the two.

If the terms of the Directive were applied solely on the basis of removal of the source of the nitrate surplus, then livestock production in the Netherlands would have to be reduced by 84 per cent; this figure could be reduced to 65 per cent if there were also to be a 28 per-cent cut in the amount of fertiliser used on crops. One radical change that could go a long way towards solving the nitrate surplus problem would be to use the animal effluent to replace inorganic sources of nitrogen. It is inevitable that because 'artificial' nitrogen is cheap while animal manure is very costly to process and move, some form of subsidy would be needed to enable such a change to take place.

Assuming the terms of the Nitrates Directive were to be applied without modification, then the impact on the output of the EU livestock sector could be significant:[18] a 12 per-cent reduction in pig production, a 10 per-cent fall in egg and broiler production and an 8 per-cent fall in milk production. The beef sector would be least affected: a fall of 5 per-cent. Sheep output would probably rise as these 'low nitrate output' animals could replace cattle in some areas.

The effect would clearly be more serious for individual countries than these aggregate figures would suggest. Belgium and the Netherlands would have to reduce their livestock numbers by 28 and 84 per cent respectively, although the Netherlands could moderate the reduction in livestock numbers by a larger cut in fertiliser use. By reducing its fertiliser use by 28 per cent it would 'only' have to cut its livestock numbers by 65 per cent. Clearly, countries like the Netherlands and Belgium are going to have to take radical measures if their livestock industries are not to be seriously damaged.

What effect is the Nitrates Directive likely to have on the Union's production and agricultural trade? The Directive has been calculated to reduce EU exports of livestock products and to increase exports of cereals, with the reduced demand for animal feed increasing the exportable surplus of grain.[19] The USDA calculates that the Nitrates Directive, combined with CAP reform, has the effect of changing the Union from a net exporter to a net importer of all livestock products, except beef. Net exports of wheat and coarse grains are estimated to increase by 10 and 50 per cent respectively.

EU countries with the most serious problems have not waited for the ponderous legislative processes of the Union to restrict farming polluters. Holland's National Plan[20] for cleaning up the environment is already in its second five-year phase. The overall objective is a 90 per cent reduction in all pollution by the year 2000. Agriculture is likely to bear the brunt of this campaign.

Already, a quota has been imposed on the numbers of pigs and other livestock that may be kept in the Netherlands. This quota can be expected to be squeezed further during the next twenty years in order to achieve a 90 per-cent reduction in nitrate and phosphate effluent disposal target. In addition, the plan proposes for agriculture: a 70 per-cent reduction in ammonia emission, a balance in the use of nitrate and phosphate fertiliser – through the use of physical limits and levies on fertiliser, a 50 per-cent cut in the use of pesticides by 2000, the construction of manure processing and disposal plants (200-million-tonne capacity by 2000) and intensification of research into 'sustainable agricultural methods'. The Dutch muck-drying scheme mentioned at the beginning of this chapter is part of this plan.

Pesticides have both a direct and an indirectly harmful effect on the environment through wide-ranging harmful impact on birds, insects and ground and soil creatures. This must be considered in combination with their impact on the habitat through the elimination of some plant species and the artificially created dominance of others. The widespread destruction of grain-eating birds and birds of prey by the over-use of organochlorine-based pesticides in the post-war decades is well documented. Though the use of these chemicals is now banned and the affected populations have recovered, it is generally accepted that the continuing uses of chemicals with a too wide spectrum of fatality to insect species and the over-use of crop chemicals is still responsible for unnecessary destruction of insects, plants and birds.

The impact of pesticides on human health is very much more difficult to measure. It is doubtful whether, apart from accidents and wilful misuse, agricultural crop chemicals pose any serious threat to human life.[21] According to Conway and Pretty (see note 4):

> Pesticide residues in average diets in the industrialised countries are all well below Acceptable Daily Intakes (ADI). Only in the case of dieldrin in the 1960s were average intakes close to ADI. Levels of organophosphate and other newer pesticides are at least 100 times less than the ADI.

Risks however do remain, with some organophosphate materials, as demonstrated by recent cases of poisoning of stockmen who have been exposed to excessively high levels of sheep-dip contamination.

But concern about water quality and possible contamination of food by pesticides is only part of the conflict between the environmentalists and farming. The 1990s will see increasing concern by urban voters over the shape and management of the countryside.

The European Commission frequently presses the EU member-state governments to consider environmental factors in the formulation and application of agriculture policy. Its Environmental Directorate (DGXI) has long had a three-point scheme for the rural environment[22] which, if it were ever to be adopted by the Council of Ministers, would place serious limits on the quantities of all types of fertilisers which farmers would be able to use, restrict the use of pesticides and place a minimum percentage of the European Community's land area under environmental management contracts. The scheme is seen as part of a plan to achieve economically 'sustainable development'.

Pointing out that agriculture and forestry control 80 per cent of the land area of the Community and thus have a major responsibility for environmental care, the Commission's environmental experts argue that modern trends in farming practice pose threats to the sustainability of agriculture itself and 'have enormous impact on other environmental and economic areas'.

The Commission's view – or certainly that of the Environment Directorate General – of agriculture's current role in the environment was made abundantly clear before the initiation of the 1992 CAP reforms.[23] The CAP, according to former EU Environment Commissioner, Carlo Ripa di Meana, is an 'ecological failure', and the damage it is doing is on a par with the environmental damage done by the polluting industries of eastern Europe. There is traditional and continuing conflict between the Environment and the Agriculture Directorates of the Commission on the environmental implications of current agriculture policy. Agricultural developments and policies should be subject to the same 'environmental impact' assessments as other industrial and public works developments, maintain the Brussels environmental specialists.

The main elements of the Commission's (DGXI) agro-environmental recommendations are:

1 maintenance of the basic natural processes indispensable to sustainable development, notably through conservation of water, soil and genetic resources;
2 decrease in the input of chemicals to the point that none of these processes can be affected in order to reach an equilibrium between input of nutrients and the absorption capacity of soils and plants;
3 achievement of a degree of rural environment management which would permit the maintenance of 'biodiversity' and natural habitats and the minimising of risks of natural disasters such as erosion, avalanches and fires. The recommendations set targets for the achievement of these objectives by the year 2000 and set out methods for their attainment.

However this conflict is resolved, it is certain that agriculture will be more and more restricted in the way it manages and maintains woodland, hedgerows and watercourses. There are likely to be greater restrictions on the construction of farm buildings and in the management of the land itself. On their own, these developments would increase rural depopulation, thus creating the paradox that it could become impossible for the European countryside to be managed in the way the urban majority would like because there will no longer be the people to work it.

This paradox has been recognised by the legislators. The Union already has policies which pay farmers not to farm. Early versions of the EU's so-called 'set-aside' policy, for example, paid any farmer who agreed to idle at least 20 per cent of his land a subsidy of up to 600 ECU a hectare to keep that land out of production for five years. The Union is also developing what are known as 'extensification policies' – paying farmers to cut fertiliser inputs and the numbers of livestock on a given area of land. These have been further extended in the May 1992 reforms.

Critics argue that these policies do not go far enough: they are concerned more with the better financial management of the CAP – by reducing production, surplus and therefore budgetary expenditure – than with better management of the countryside. The conservationists want to see the subsidies being actively directed towards the replanting of hedges and woodland, to the preservation of wetlands, and the maintenance of 'wild' areas which will sustain habitats for Europe's threatened flora and fauna.

The cost of a more environmentally 'biased' policy such as this might be thought to be excessively expensive. It is, however, probable that it would be no greater than the cost of the present policy. The UK Country Landowners Association (CLA) has calculated[24] that the cost of such a policy, where the current production-orientated supports and subsidies were replaced by payments designed to pay farmers to 'manage the rural environment' and with alternative sources of income in rural areas, would be no more than the current

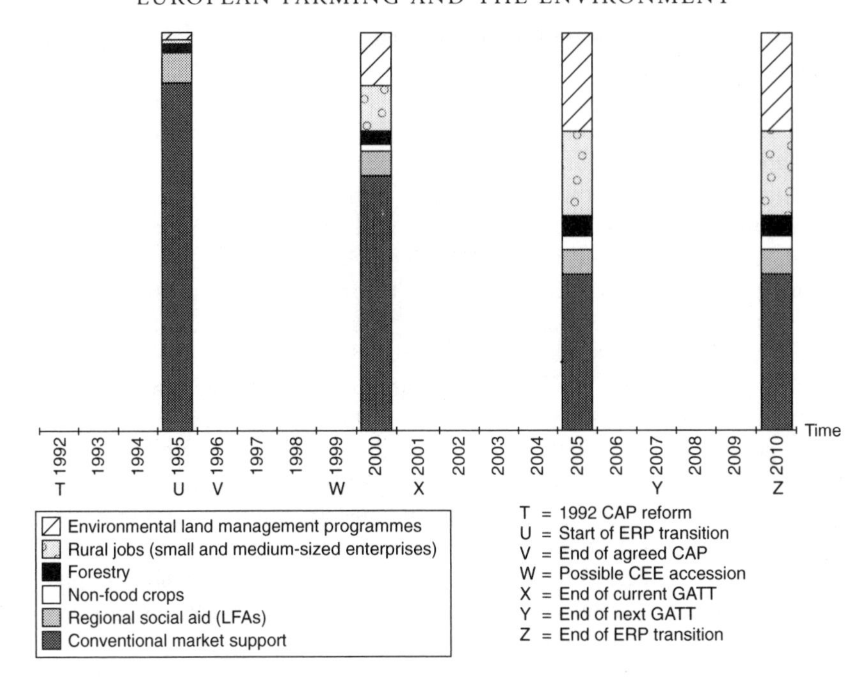

Figure 8.3 Changing the emphasis of rural support policies from agriculture policy to a European Rural Policy (ERP), as seen by the UK Country Landowners' Association. While the total amount spent may not change, the proportion spent on support of production declines is to be replaced by spending on environment and rural development

Source: Focus on the CAP, CLA, London, 1994

UK share of CAP costs of approximately £3 billion (8 per cent of 37 billion ECU). The CLA freely admits that most of the money currently spent on agricultural support is socially wasteful since it merely translates into inflated prices for land, machinery and variable agricultural inputs. Why not use the same money for what is perceived by the majority of people as being more socially desirable: environmentally acceptable management of the countryside?

If the true cost of the United Kingdom's share of the cost of even the 'reformed' CAP is taken into account, such a policy would in fact be cheaper than the only slightly modified CAP. This is because the Brussels' budget figure does not include the 30 to 40 billion ECU which EU consumers have to pay for import protection. On a typical arable farm of 142 hectares, the removal of production-orientated subsidies and the reduction in output resulting from more environmentally friendly management would result in a 29 per cent reduction in the farmer's income. The actual subsidy required to maintain income at the original level would however be cut by almost three-quarters – for the simple reason that input costs under less intensive production methods will also have declined substantially.

The scientific evidence would suggest that the conservationist's concern about habitat destruction by modern agriculture is fully justified. In west Germany the 'Red List' of endangered species shows that 30 per cent of flowering plants and fern species, 40 per cent of bird species, 50 per cent of mammal species and 65 per cent of crawling species and amphibians are either extinct or endangered. Scientific evidence points overwhelmingly to agriculture as the major destroyer of these species. Of the 700 endangered plants on the German list, 75 per cent are said by scientists to be threatened by agriculture.[25]

It is not only the scale of the potential destruction by modern farming methods, but also the acceleration in the speed of destruction which is so threatening. Another West German study of 933 species of plants showed that in the period 1870 to 1950, fourteen had disappeared; in the following three decades 130 became extinct, while a further fifty species were threatened with extinction, seventy-four were seriously endangered and 108 were in decline.

The blotting-out of plant species also generally carries with it a multiplier in terms of dependent animals which disappear when their food supply no longer exists. It is estimated that ten to twelve animal species disappear with every plant species which is eliminated.

How well will the EU's Common Agricultural Policy respond to these changes? Can it become a rural policy which meets all the demands of society for intelligent and sustainable food production and countryside management, or will it continue in what many regard as its so far inappropriate way? The current indications are that it will not change fast enough to deal with the new needs. The Union is still spending over 30 billion ECU a year which it could be said is aimed at encouraging production, while still allocating less than 5 billion ECU to the structural and social measures that would sustain communities and create a more attractive countryside. Pessimists say it will take at least another EU budget crisis and much greater pressure on the agricultural policy-makers from finance ministers, consumers and environmentalists, before the fundamental changes which are still needed will take place.

Meanwhile, national governments will be forced to introduce their own legislation. This will undoubtedly penalise agriculture for its role as a major polluter. This type of legislation will by the end of the century have a more profound effect on the pattern of EU farming policies.

The extent to which the 1992 CAP reforms will achieve an environmentally acceptable agriculture, or whether more specifically environmentally biased agriculture policies are required, has, of course, been studied by agricultural economists and environmentalists. It is probable that the consensual conclusion is that the 1992–96 reform programme on its own is unlikely significantly to reduce the use of fertilisers and other agricultural inputs. Nor will it significantly reduce the potential pollution by fertilisers to the same extent as policies specifically designed with an environmental intent.

A detailed study of this question was carried out by economists at the US Department of Agriculture in 1993–94. It looked at the extent to which the CAP reforms would be likely to achieve environmental objectives – particularly the reduction of nitrate and phosphate pollution from both arable and livestock farming – compared with specific environment policies such as the Nitrates Directive or a possible specific fertiliser tax.[26]

Comparison of five alternative policies on EU farming – (1) a 50 per cent fertiliser tax; (2) the current Nitrates Directive; (3) the 1992 CAP reform without (1) or (2); (4) CAP reform and the Nitrates Directive; and (5) a combination of fertiliser tax and the Nitrates Directive – indicated that only the last, policy (5), would result in significant pollution reduction.

CAP reform alone will result in very little reduction in the total amount of nutrients delivered to the soil – in either livestock or arable farming areas. Delivery of nitrogen and phosphates is likely to decline less than 5 per cent.

SOIL DEGRADATION

Soil degradation, manifesting itself primarily and most obviously in erosion, is one environmental problem that has been a by-product of agriculture since man first abandoned the nomadic life and became a farmer. Its effects have, however, been exacerbated by modern agriculture. Mechanisation, modern streamlined crop rotations – in some areas the abandonment altogether of alternating one crop with another to control disease and maintain fertility – have speeded up the process of destruction of the soil structure and in extreme cases the loss of the soil itself.

According to a report[27] to the European Parliament, at least 10 per cent of the Union's soil area is threatened by erosion. The problem is worse in southern, Mediterranean areas of the EU: in Portugal it is estimated that more than 20 per cent of the current crop area is threatened by erosion and ought in fact to be taken out of agricultural production. The problem is not, however, limited to the south: German agricultural extension experts reckon that on between 50 and 66 per cent of the land in Bavaria erosion is currently above a tolerable level. The EU Commission's Environment Directorate General (DGXI) estimates[28] that in the northern areas of the Union the average annual loss is more than 8 tonnes/ha. Over considerable areas the annual loss of humus could be as much as 100 tonnes/ha.

But modern farming has brought with it a more insidious loss of soil reserves: the steady wasting away of topsoils under the influence of unsympathetic farming practices. Under-drainage, cultivating under unsuitable conditions, inappropriate cropping and the removal of woodland and hedgerow windbreaks have been shown to result in annual losses in most vulnerable areas of up to 45 tonnes of topsoil a hectare.

THE GENETIC THREAT

'At present, about 50 per cent of all animal and plant species in Central Europe are endangered.' This quotation from a leading expert on agriculture's role in the European environment[29] is the most damning condemnation of man's role in destroying the biological richness of the planet. Agriculture is, of course, not wholly to blame for this potential genocide. It cannot, however, escape the accusation of being the major culprit responsible for the loss of many important species of plant and animal; it is also said to be squandering by modern breeding methods the genetic reserves of its own industry by eliminating genotypes of useful farm animals and plants that had been accumulated over the centuries. To take one example: most of the many native regional breeds of cattle that once predominated in the farmyards of western Europe have been displaced by the highly bred – many would say over-bred – Holsteins and Friesians that now form 90 per cent of the EU dairy herd.

Governments and political parties ignore the environmental issue only at their peril. The problem for the politicians is, however, to reconcile the environmental requirements of society with its desire for continuing economic growth, control of government spending and the control of inflation. Politicians also have to face the reality that satisfying the environmental desires of the majority is essentially a zero-sum game: cleaner rivers, lakes and seas as well as uncontaminated water, a pleasant countryside, the preservation of animal and plant species – all these highly desirable objectives can be achieved only if someone is prepared to pay.

Though it has yet to be spelled out to the people, a price is having to be put upon these 'environmental goods' for which, ultimately, the consumer is going to have to pay either through higher prices for food and other goods or through taxation.[30] The replacement of agricultural market policies with policies designed to preserve rural communities on the basis of extensive rather than intensive agriculture will cost money in terms of subsidies from the public purse. Initially, the imposition of environmental controls on agriculture will mean that the farmers will be worse off.

Eventually, they will be partially compensated by higher prices for food which may become scarcer as a result of production limitations. What is not clear is how far the agricultural polluter will be expected to pay, through reduced output, without compensation from national or EU exchequer funds. Certainly the overtly 'structural' type of environmental policy will mean subsidisation by the taxpayer. The grey area is, however, the extent to which loss resulting from direct environmental controls such as limits on fertiliser use and animal effluent disposal will be compensated. Only if means of increasing economic growth can be found which do not further damage the environment and thus generate increased wealth to pay for these environmental 'goods' can society improve the environment without overall loss and

Box 8.10 The European Union's environmental policies

The EU's initial common approach to the environmental problem grew out of the United Nations Conference on the Human Environment held in Stockholm in 1972. The major preoccupation of that conference was with global action on pollution, soil degradation, resource depletion, ecological damage and climatic modification. It will be noted that agriculture plays an important part in every one of these five main headings. The Union's first Action Programme on the environment was adopted by the EU Council of Ministers in November 1973. It forms the basis for all of the Union's subsequent legislative activity in this area. It laid down six main principles: prevention of potential environmental threats at source, adoption of the principle that polluters must pay, that the Union and the member states must collaborate in international action, that policy should coordinate national and EU level policy developments, that there should be compatibility with economic and social developments, and that EU policy must take account of developing countries in its environmental policy.

Four main fields of action were laid down: reduction of pollutants and other nuisances, non-damaging use and rational management of land, environment and natural resources, general environmental protection and international cooperation.

The original EU action programme was subsequently updated by the Second Action Programme in 1977 and the Third Action Programme in 1983. Environmental policy became a major policy area under the Revision of the Treaty of Rome (the so-called 'Single Act') in 1986. The Fourth Action Programme was subsequently agreed in 1987. Article 100 A (3) of this revised Treaty says: 'The Commission in its proposal . . . concerning health, safety, environmental protection and consumer protection will take as a base a high level of protection.'

Particular focus was put upon the agricultural aspects of Environment policy by the EU Commission's 1988 paper, 'Environment and Agriculture'. From these various policy initiatives have arisen the subsequent adoption by Council of some 120 Directives, Regulations and Decisions in the environmental area.

particular loss for one section of society – in this case, the agricultural industry. Biotechnology may provide the key to this conundrum, but this too brings its own risks and costs with it.

PROTECTING SPECIES

Preservation of animal and plant species was a major objective of the Union's first Action Programme for the Environment, established in 1973. This has been strengthened in the subsequent programmes. The main aim is to achieve an integrated approach to the conservation of endangered species of flora and fauna and the protection of habitats. An important feature of the Second Programmes was the initiation of a programme for the protection of birds, the control and restriction of international trade in certain species of wildlife, the re-assessment of shooting and hunting laws and the protection of wetlands.

One of the most important pieces of legislation in this area is the Directive on the Conservation of Wild Birds (1979), which sets out provisions for the protection, management and control of all species of naturally occurring birds in the wild state. Also of high importance is the Union's endorsement of the Berne Convention on the Conservation of European Wildlife and Natural Habitats. In addition to listing protected species, this Council decision calls on all member states to foster policies which conserve wildlife species. In 1984 this decision was backed up by a Council Regulation granting financial aid to environmental projects specifically aimed at preserving wildlife habitats, in particular the biotopes most likely to harbour endangered species.

RECONCILIATION OF ENVIRONMENTAL AND AGRICULTURAL POLICIES

The need to sustain modern agriculture in developed countries, both to maintain food supplies and to conserve the countryside itself, is generally accepted by major environmental groups. The means of reconciling the too frequent conflict of these aims of agriculture and rural policy are embodied in the World Conservation Strategy – 'Caring for the Earth – a Strategy for Sustainable Living,[31] evolved by the World Conservation Union, the United Nations Environment Programme and the World Wide Fund for Nature in 1991.

Its main message was that further economic development is necessary to improve the living conditions of the billion or more people who currently live in poverty and to meet the needs of the 3 to 5 billion more people who will be added to world population during the next fifty years – but that in achieving it the vitality and diversity of the Earth must be conserved.

The more realistic conservation organisations, rather than rejecting modern agriculture out of hand, accept the need for compatibility of a viable agriculture with preservation of habitats. In a statement on agriculture and conservation,[32] the UK branch of the World Wide Fund and the Royal Society for the Protection of Birds recommend the maintenance of a prosperous rural economy, producing sufficient quantities of high-quality food for the nutritional needs of Europe and contributing to an equitable trading system; an integration of land uses; a sustainable agriculture which internalises the environmental and social costs of its activities, applies the 'polluter pays' principle and is subsidised only for achieving wider social and environmental objectives rather than for food production in itself; and a wildlife-rich countryside within a diverse, prosperous and peopled landscape.

THE ENVIRONMENT AND AGRICULTURAL TRADE

Finally, there is the highly contentious issue of the relationship between agricultural trade and environmental stress. The argument lies between those who argue that trade liberalisation – even of the limited nature envisaged under the 1993/94 Uruguay Round agreement – leads to greater pressure on the environment from agriculture and those who believe that it leads to less.[33]

Followers of the second of these two schools of thought argue that agriculture policies which distort trade are also likely to be largely responsible for damaging the environment. Although it does not necessarily follow that eliminating these protectionist policies will automatically lead to more environmentally acceptable agriculture policies, liberalisation should provide the basis for establishment of policies decoupled from market support, but firmly linked to environmental protection policies. This is increasingly the view of leading agricultural economists, who see the prospect of an eventual agreement on world agricultural trade liberalisation as a golden opportunity for the advancement of new agriculture policies which take account of environmental costs.

To prevent trade distortion – and to prevent national governments using environmental legislation as a non-tariff barrier – it is inevitable that there has to be international organisation of agro-environmental policies. Consensus agreements on environmentally desirable policies will become as important as agreements on the rest of agricultural trade.

The alternative, essentially protectionist view of the impact of liberalisation is that the GATT agreement will be bad for the environment. The removal of frontier protection of agriculture, the argument goes, would prevent governments from maintaining the sort of family farm-based agriculture, which it is automatically assumed is 'environment-friendly', and open up the prospect of a world food trade dominated by the multinationals.

Apart from its illogicality, this view embodies more than one unfounded assumption. Most obviously, there is no evidence that trade protection has in fact been used by governments to conserve a particular type of farm structure. In the European Union, for example, the massive protection and subsidisation of agriculture of the last twenty-five years did nothing to prevent the drift from the land and little to prop up the incomes of the type of farms that are considered by the green faction to be desirable. Intensification continued apace in this period and the average farm size continued to increase, with many small farmers leaving the land as their farms were amalgamated into larger holdings.

Much the same pattern can be observed in other developed countries – except in those where governments have introduced separate policies, additional to market protection, in order to preserve a particular type of farm structure. The Scandinavian countries are an outstanding example of this deliberate policy of fostering a particular type of farm structure.

The Nordic countries have used clearly defined social and structural policies to maintain the population of remote northern rural areas. The fallacy of the extreme green argument against the Uruguay Round is to assume that policies which protect the rural environment could not be maintained in the context of a set of international rules governing trade in agricultural commodities and the national policies underlying that trade. The truth is that the GATT can have little influence over policies which are de-coupled from support of agricultural markets – which is exactly what effective agro-environmental policies must be.

If the propensity of agriculture to pollute soil, habitats and water supplies is to be controlled, farming has to be brought into the framework of rules that governments are very slowly evolving to make all polluters pay for the damage they are doing or might do to the environment. Liberalisation of farm trade will not affect this process. What the GATT will do, however, is to establish a set of internationally acceptable rules on agricultural environment policies which prevent governments from using their environmental legislation as a non-tariff barrier behind which they may continue protecting their agricultural industries.

The agricultural industry's potential to pollute is indisputable. The important question for the future is: does agriculture need to pollute to maintain adequate food supplies? What is not in dispute is that agriculture bears responsibility for a large part of the ecosystem and is therefore most responsible for the biophysical impact feedback to which Professor Pearce[34] refers.[35] Agriculture, by its very nature, seeks quite deliberately to change the ecosystem. In an agricultural context sustainability is the ability to maintain productivity when subject to stress or shock. Stress is defined as a small but steady change in the physical production conditions, while shock is sudden change like drought or exceptional rainfall, the advent of a new plant disease or a change in the financial conditions under which agriculture operates.

According to Pearce:

> The unchecked abuse of resources within an agro-ecosystem, whether as a result of the inappropriate use of agrochemicals and fertilisers, the over-cropping of erodible soils, poor drainage, etc., not only directly affects sustainability of the agro-ecosystem but may also increase its susceptibility to other external stresses and shocks, such as changes in market conditions, prolonged dry seasons, changes in land tenure and so on.

There is little doubt that systems of farming that do not pose environmental challenges are economically feasible; there is also little doubt that under current policies and price relationships they are less profitable than systems employing all the most profitable techniques currently available.

Many environmentalists would go to the point of banning the use of all chemicals and all intensive methods of livestock husbandry. Sustainable though such an agriculture may be in ecological terms, how far would it be capable of maintaining food supplies?

The major problem that has to be faced is that modern food production is highly dependent upon fertiliser produced outside the agriculture industry itself. In modern European agriculture, it is estimated that 60 per cent of total output results solely from the application of chemical fertiliser alone. Even to halve this input would not only result in substantial reductions in output but also in bankrupt farmers. Estimates by the British Agriculture and Food Research Council[36] indicate that on a typical arable, mainly cereal farm, a 30 per-cent reduction in the use of nitrogen on crops, while reducing wheat yields by 22 per cent, would reduce the farmers' profit by 112 per cent – in other words, put him into a financially unsustainable position.

To take an extreme view: a complete ban on the use of nitrogen in agriculture would require the farmed land area of the United Kingdom to increase by nearly 150 per cent – clearly a physical impossibility – if current levels of production were to be maintained. Of course, current levels of British food production do not need to be maintained: considerable economic advantage could be gained by reducing UK arable output by at least a third and importing the deficit. This increased import requirement would, however, have to come from non-EU sources if the environmental advantage were to be maintained (there could also be the danger of 'exporting' the environmental cost of maintaining European food supplies). None the less, a complete nitrogen ban would create an unacceptable problem for the agricultural industry. For Europe in general the result would be similar.

Given that present methods of intensive agriculture cannot be continued unmodified, what are the alternatives? In 1989 the US Academy of Sciences set up a Committee on the Role of Alternative Farming Methods in Modern Production Agriculture. The report of that committee[37] – which is as relevant in Europe as in North America – was unequivocal about the potential for environmental harm of modern farming methods. It concluded that there is an alternative.

The broad conclusions on alternative farming are clear: farmers are able to reduce their pesticide use on commercial cereal crops through rotations that disrupt the reproductive cycle, habitat and food supply of major crop insect pests and diseases. Alteration of the timing and placement of nitrogen fertilisers will often allow farmers to reduce application rates with little or no sacrifice of crop yield.[38]

Further reductions in fertiliser use are possible in regions where leguminous forages and cover crops can be profitably grown in rotation with major arable crops. Commercial fruit and vegetable growers in many situations can dramatically decrease pesticide use with an integrated pest management (IPM) programme. Sub-therapeutic use of antibiotics can be reduced or eliminated without sacrificing profit in most beef and pig production systems which are not based on 'extreme confinement' rearing. To adopt a more environmentally acceptable style of farming, without going as far as all-organic, would undoubtedly involve a considerable reorganisation of the agricultural industry.

On an international scale, the mathematics of organic farming are relatively simple: adoption of an all-organic policy would lead to a large world food deficit and would seriously cripple the ability of the agricultural industry to feed an increasing population in the future.

9

THE 'BIOTECH' REVOLUTION

Producing more from less

Biotechnology:

> The application of biological organisms, systems and processes based on scientific and engineering principles, to the production of goods and services for the benefit of man.
>
> OECD 1982

In 1980 American scientists interfered with the cells of mice, extracted the substance which this interference produced, injected it into dairy cows and, overnight, threatened to wreck the tender balance of the world's too often over-supplied market for butter, milk powder and cheese. What those scientists were doing was for agriculture as epoch-making as in a much wider way the splitting of the atom was several decades earlier. The application of DNA technology to the manipulation of animal and plant life, the basis of what has become the modern science of 'biotechnology', offers to more than double post-war gains in productivity in cereal-growing, milk production and the production of pork, lamb and beef – and to do it much faster than was ever possible before.[1]

It also offers the means of manipulating plant genes so that not only the amount but also the type of product can be produced to order. But so, of course, could the more traditional animal- and plant-breeding techniques – though much more slowly. Under more traditional methods the numbers of generations that would be necessary to achieve the breeder's objectives would take years – and in the case of large, slow-gestating animals like dairy and beef cows, decades.

The techniques of DNA manipulation and allied, more manipulative processes like ovum transplants have allowed scientists to concertina the process of animal and plant improvement into a fraction of the time that it once took. The regular injection of dairy cows with what is now known as 'bovine somatotropin' (bST) – the product of the mouse's manipulated cells – is one of the more spectacular products of the modern science of biotechnology. The bovine growth hormone, bGH, is naturally secreted by the animal's pituitary gland. In the early 1980s, scientists discovered how to isolate the bGH-producing gene and to transfer it to bacterial cells; this allowed the mass production of the 'recombinant' bovine somatotropin or genetically engineered hormone.[2] Regular injections of 30 mg into each cow on a regular basis produced spectacular increases in milk yield – under experimental conditions, beyond 20 per cent.

Under commercial farm conditions, bST has the effect of increasing the milk of dairy cows by between 12 and 15 per cent – instantaneously. The technical and more – importantly – the political and social implications of this innovation are enormous. In the United States, it is estimated[3] that it would have been possible in 1990 to produce that year's 30 million tonnes of milk with 11 per cent fewer cows, with 2.5 million tonnes less grain feed and 6 million tonnes less effluent to be disposed of.

In the European Community, the political implications are even more dramatic. Were the use of bST to be allowed, the Community's milk production would have been increased by approximately 15 million tonnes.[4] This would have effectively doubled the EC's already large milk surplus and almost doubled the enormous 5 billion ECU annual farm budget commitment for supporting the dairy industry. Conversely, because such an increase in output would not be allowed under the Community's rigid quota system, the number of cows needed to produce the allotted quantity would have been reduced by around 11 per cent or about 2 million animals. Since large farmers are better able to adopt new biotech techniques, this would have meant that large numbers of Europe's small farmers would have been put out of business, it is argued by the European Commission. This, along with consumer fears about the effects on human beings of the use of bST, is the major reason why the use of the substance on EU dairy farms is now – at least temporarily – banned. A twelve-month moratorium on the use of bST, first introduced in 1989, has been extended every year since, while the politicians decide what their position should be on this potentially explosive issue and determine what action they should take. EU agriculture ministers have no doubt about the potential threat to the equilibrium of the dairy market: they have extended the initial moratorium on its use until the end of the current dairy quota policy in 2000.

The use of recombinant somatotropin could be equally startling in other farm livestock. Experimental work with porcine somatotropin (pST) shows not only that the same amount of pork or bacon can be produced with 20 per cent less feed, but more important to modern dietary standards, the amount of fat in a pig carcass can be reduced by as much as 80 per cent. Probably as important for the Europeans with their large numbers of intensive livestock crammed onto too few heavily inhabited hectares, the amount of manure produced by pigs being treated with porcine somatotropin is cut by more than 15 per cent. Work on the use of somatotropin on sheep – where the effluent problem is not present – still offers the prospect of substantial meat production and quality gains. Increases in production of as much as

36 per cent and reductions in fat content of the meat of 30 per cent have been achieved under experimental conditions.

Biotechnology – the manipulation of natural processes to provide useful organic products – is in a basic sense not a 'new' concept.[5] Brewing and bread-making are early examples of human manipulation of naturally occurring biological processes to serve an economic objective, but the modern use of the term and the industry which it has created result largely from the successful investigation by scientists of the nature and role of the nucleic acids DNA and RNA. This permitted man to do two things: to unravel the 'genetic code' at the source of all animal life and to develop 'recombinant DNA' technology. This allowed scientists to achieve two vitally important objectives: not only to be able to mass-produce genes in bacteria, but also to transfer fabricated gene constructs to other organisms. This enabled the incorporation in animal and plant organisms of valuable characteristics which would not have been possible with conventional breeding methods. Genetic engineering allows breeders to leap the gap between slow, chancy conventional breeding methods and gene transplantation, where the desired combination of genetic characteristics can be incorporated into one animal or plant. The desired individual can then be replicated by cloning.

In crop production, biotechnology is currently being used for the rapid identification of valuable genes, for the development of new methods of hybridisation, for plant propagation and tissue culture, for breeding herbicide and insect tolerance, for altering the efficiency with which the plant utilises moisture, nutrients and the energy of the sun and resists disease. One of the potentially most fruitful developments from both a plant productivity and an environmental point of view is the development of 'biopesticides'.[6]

In animal husbandry, the advantages of biotechnological techniques are even more promising because the generational obstacle to improvement is much greater than in plants. To breed a new strain of dairy cow or bacon pig would until recently have taken a great deal longer than the time needed to breed a new strain of wheat or sugar beet. The use of bovine somatotropin is an example – though not involving breeding – where the biotech innovation gives productivity increases far greater than those accruing from earlier breeding techniques. Whereas the application of the 'best strains' of milking cow genes to dairy herds in the United States is estimated to have added 50 kg of milk a year to average yields, the use of bST adds 1000kg.

The main current objectives of the application of biotechnological techniques to animal husbandry are: the improvement of quality; the improvement of animal health through the rapid breeding in of disease-resistance; and the improvement of animal welfare through the detection of management stress effects. Recombinant DNA technology, the use of monoclonal antibodies, cell fusion and protein engineering are all currently being used to improve the quality, health and welfare of farm livestock. An area of considerable importance where commercial techniques are already available is the monitoring of animal health and the control of disease. A good example of the use of the recombinant DNA technique is the production of vaccines from an antigenic sub-unit of the potentially harmful microbe which stimulates the animals' immune system to resist diseases, but without any pathogenic effect.

The improvement of feed efficiency through the provision of feed additives containing vitamins, enzymes and amino acids is another fruitful area for commercial development. Embryo multiplication and gene transfer are techniques already being employed by specialist livestock breeders. Another area regarded as likely to be highly fruitful in the future is the 'transgenic' production from farm animals of products for use in human medicine; typical is a blood product for treating haemophiliacs which is produced by farm animals.

Although considerable effort and resources are currently being devoted to means of increasing agricultural production by biotechnological methods, it is generally agreed by the companies involved and by independent researchers that significant effects on agricultural production – both in crops and livestock husbandry – cannot be expected until the late 1990s or even the first decade of the next century. Major effects cannot be expected until the first and second decades of the next century. None the less, it is likely that 'conventional' plant and animal breeding as well as improved husbandry and management will tend to continue to increase production at a steady rate through the second half of the 1990s.

In those countries where governments are most sympathetic to biotechnology – the United States, Canada and Australia – there has been a dramatic increase in the number of agricultural biotech patent registrations.[7]

What is certain is that food production inside and outside the European Union will increasingly be able to keep up with world population growth due to the ability of science to overcome current agricultural problems – particularly the environmental constraints already pressing on the agricultural industry.

With biotech assistance crop plants can be bred to produce more of their own nitrogen, plant management systems can be developed allowing biological control of insect pests and diseases, somatotropins can be used in meat production to allow more leaner meat to be produced with less feed input and less effluent output, and the general use of genetic engineering allows more food to be produced with less inputs.

World food production could easily be doubled in the next forty to fifty years, it is currently estimated,[8] as a result of these new technologies.

Major plant-breeding innovations which are likely to make an early contribution to continued agricultural productivity and production increases include:

1 the development of high-yielding winter wheats with high-protein/high-lysine contents;
2 the development of high-yielding white maize hybrids for use in Africa;

3 the development of high-yield, high-protein triticales for animal-feeding in Europe;
4 the development of high-lysine, high-oil, high-starch maize varieties for the production of food, fuel, chemicals and industrial feedstock;
5 the development of new, high-yielding, double-zero rapeseed varieties which will grow in northern climes;
6 the production of new soya varieties which will grow closer to the Equator;
7 the expressing of nitrogen-fixing genes in non-leguminous crops so that dependence on added nitrates could be substantially reduced; this is likely to be a reality by the year 2000.

The breeding of plants for tolerance of environmental stress is nothing new: considerable research is currently being conducted into the development of plants which are capable of withstanding drought, frost, heat and salinity. Considerable progress could be made in these areas in the course of the next twenty years.

Dramatically increased productivity and quality of crop plants is, however, likely to come from genetically engineered plants which are resistant to herbicides, disease and insects. Biotechnology is also expected to provide new crop protection products for dealing with weeds, insects and diseases. It is unlikely that such controls will be exclusively biological, but rather form part of integrated pest management (IPM) systems which will still employ control methods of a chemical nature. This combination will, however, lead to the use of very much less chemical and to the use of products of a more environmentally benign type than those used at present. Biotechnological controls are playing only a modest role in agriculture in the mid-1990s but will become progressively more important after the turn of the century.

Although biotech methods are estimated to occupy only 2 per cent of the US crop-protection market and an even smaller proportion of the EC market, it is likely that the 'new generation' of chemicals will be designed to fit into more biotechnical control systems and will 'have novel modes of action, extremely low use rates, high selectivity for target pests, high safety for mammals and non-target species, high environmental compatibility and excellent cost efficiency'.[9]

Disease and insect control by biological means is already well advanced at the laboratory and field-trial stage. In particular, recombinant DNA techniques have been used to produce virus resistant plants; 'cross-protection' or the inoculation of plants against virus-diseases has already been used on a commercial scale.

What is clear is that advances in biotechnology capable of being used in commercial crop production have taken place much faster than scientists and economists originally expected. Predictions made in the later 1980s that biotech methods would not be applied to farm-scale crop production until well into the first decade of the next century are firmly contradicted in a comprehensive report on biotechnology in agriculture published by the US Congress Office of Technology Assessment.[10] 'Research in crop agriculture has advanced at a much faster pace than anticipated just a few years ago,' says the report.

Modified microbial insecticides as well as genetically modified plants capable of resisting major bacterial and viral diseases were already in commercial use in 1995. Bioherbicides and genetically induced herbicide tolerance were also already in use in the mid-1990s.

The estimated effects on productivity vary. Under the most likely scenario of the continuation of present trends and a 2 per cent a year increase in research spending:

> Feed efficiency in livestock production will increase at an annual rate of from 0.39% for dairy cows to 1.62% for pigs. In addition, reproduction efficiency will also increase at an annual rate ranging from 0.67% for beef cattle to 1.25% for pigs. Milk production per cow per year will increase at 3.01%, from 14,200 pounds to 19,200 pounds a cow in the 1990–2000 period.

For several of these developments the increase in productivity is, however, less than was expected in estimates drawn up seven years ago. The expected increase in commercial milk yields, for example, is in the mid-1990s less than estimated in 1985 – due to the continuing legislative barriers to the use of bST to boost milk yields. In the poultry industry, it will be extremely difficult to boost feed conversion efficiency, due to the very high level achieved over the last two decades. The largest gains are likely to be in beef and pig production. In the crop sector it is likely that productivity increases will be greater than the original 1985 OTA estimates.

Assessment of the potential advantage and economic effect of the application of biotechnology to agriculture and food production is difficult. The word 'potential' is used advisedly; the political obstacles to the exploitation of the new techniques are considerable (namely, the EU Commission and Council obstruction of the use of bST in dairy cows). Transgenic animal 'experiments', rogue genes escaping into the wild, potential food contamination are only some of the nightmares that appear to haunt the minds of the opponents of commercialisation of biotechnological innovation. The economic and social aspects of biotechnology are also highly important and pose new problems for the agricultural policy-makers – as demonstrated by the EU attitude to bST.

In theory, it should be possible for many of the environmental obstacles to the maintenance of agricultural production to be overcome by the development of new methods of crop and animal production which manipulate existing, inherent natural features of the plant and animal organisms with less recourse to 'artificial' external materials. There are two problems however: (1) the comparative slowness of development of commercially viable biotechnological solutions to problems in plant and animal husbandry set by the probable restriction on herbicides, fertiliser use and animal effluent disposal and health problems; and (2) the political obstacles to the development of these new biotechnological methods.

Some scientists, however, earnestly believe that though environmental pollution, soil erosion, genetic inflexibility and global warming may well pose potential threats to the maintenance of adequate world food supplies, they can be mastered – in time – by science. According to this optimistic assessment, science is quite capable of overcoming these problems now and in the future; world agriculture will continue to increase production faster than the increase in the number of mouths to be fed.

Convinced that new developments in agricultural science are likely to overcome undoubted obstacles to further expansion of food production is John R. Campbell of the University of Florida.[11] He says:

> With the continued proliferation of advanced agricultural technologies and management practices, it seems likely that global agriculture will meet and exceed the tremendous challenge of doubling food production within the next 40 to 50 years. With the added impetus of biotechnology and other new scientific tools, we see clear indications that many of the problems and constraints of the 1960s and 1970s have been surmounted. Given adequate farmer incentives, favourable government policies and the essential marketing transportation and credit infrastructures in developing and developed countries, the global agriculture and food system seems poised for significant new growth as we approach the 21st century.

The unequivocal message is that the analyses of more pessimistic agricultural futurologists – such as Lester Brown of the Washington Worldwatch Institute[12] – that food production will increasingly not be able to keep up with population expansion, take little account of the ability of science to overcome current agricultural problems.

The US National Research Council report lists four main areas of technological development which will increase animal productivity:

1 the use of recombinant peptide hormones and other growth enhancers in livestock;
2 advanced cellular engineering techniques that will fundamentally change the basis of animal reproduction;
3 direct gene transfer in animals to develop totally new and dramatically improved strains of transgenic livestock;
4 gene transfer to develop a system of 'molecular farming' which would allow the production of a wide variety of non-food protein (like, for example medical products such as insulin, growth hormones, interferons and so on).

Ova implantation, like artificial insemination, is one of those techniques that allow farmers to obtain more for less: more beef and more milk from fewer cows means fewer inputs – less nitrate and other effluents and less work to produce a given amount of meat or milk. The techniques involved here are part of the sort of biotechnological development which is going to become increasingly important in the 1990s.

In the livestock sector there are three major areas which are likely to play an increasingly important role in improving agricultural productivity in the 1990s: embryo technology, gene technology and hormone and immunotechnology. The US Office of Technology estimates that during the next twenty years the world's agricultural output will increase at 1.8 per cent a year. This growth will come principally from increased yields and very little from greater land areas devoted to farming. Increasingly, this rise in output will have to be achieved with less environmental damage; such a result can only be achieved with less conventional inputs and more productivity-raising, but environmentally benign, innovations. The need to establish a legislative framework that does not obstruct such developments is clear when it is realised how soon many of the new scientific developments will actually be in commercial use.

It should be noted that productivity increases without the application of biotechnology are only modest. It is therefore clear that, given the limitations which are likely to be placed on developed world agriculture, new technology will need to be applied if production is to be maintained. It follows also that the EU agriculture industry needs to be allowed to utilise the scientific innovations which will become available if it is to maintain what society in general will come to regard as a necessary level of production and an environmentally acceptable agriculture industry.

The question is, however: how far will scientists, commercial organisations and farmers be allowed to develop these new methods without interference from the politicians? The EU experience with bST is not encouraging. The EU Commission has now put off a decision on the approval, or otherwise, of this product three times since 1989, and it is clearly continuing to play for time while it considers the political aspects of the decision.

The EU's Committee for Veterinary Medicinal Products (CVMP) in 1991 recommended that bST be cleared for commercial use. The Committee was given the task of clearing bST on human health and animal welfare grounds after the EC Council of Ministers decision in December 1990 to extend the moratorium on bST for a further year, until the end of 1991. Despite the scientific assurances, however, in December 1994 the Council further prolonged the bST moratorium to the end of the century.

The committee's role was to assess whether new veterinary products meet the traditional EC criteria of safety, quality and efficacy. This it did. However, in the specific case of bST the Commission continues to worry over, not its human health or animal welfare impact, but the product's likely socio-economic impact if generally used by the European dairy industry. This so-called 'fourth hurdle' which new biotech products will have to clear is going to become increasingly important in the development of productivity-boosting innovations. In the case of bST, the fourth hurdle to be overcome before its use can be legalised – economic and social criteria – concerns mainly the possible effect of the use of the hormone on consumers' perception of the safety of milk and dairy products. Its commercial use is not considered likely to enhance the attractiveness of a product for which there is already declining demand. Similarly, it is considered by Commission officials that the use of a product which will increase milk yields by 10 to 15 per cent will have serious structural and political implications for the dairy farming sector. These fourth hurdle criteria are thus nothing to do with the actual scientifically assessed safety criteria.

Because of its exceptional economic, social and political aspects the bST issue is likely to prove a test case establishing whether or not the fourth hurdle may or may not be employed in the Community's regulation of the use of biotechnological products. The Council decision to prolong the prohibition on bST use confirms that it will be. It also raises the important question of whether or not the Commission is the right body to have responsibility for this work. There is an influential school of thought that continues to argue for the establishment of a politically independent authority within the EU along the lines of the US Food and Drugs Administration to handle these matters.

Two broad political questions that dominate the adoption of biotech: will it alleviate or aggravate the problems of modernising and adapting agriculture in developed countries to more liberal protection and support policies; and will it not exacerbate the problems of over-production and surplus? The OECD highlights the dilemma:[13]

> It has been argued that biotechnology's big potential should be used rather for cost reduction than for yield increases in order to reduce both agricultural surpluses and environmental damage. However, this strict distinction is hardly ever possible: to be adopted, any new technology must be profitable. Any increase in profitability will induce farmers to extend production, unless quota systems restrict them.

The economic effect of biotechnology could, if political conditions allow, be most dramatic in the livestock sector; but this is also the sector where 'interference' from consumerist, animal welfare and environmental groups is likely to be greatest. According to the OECD's estimates, productivity in the livestock industry could rise with the application of recombinant somatotropins by 4.3 per cent a year, compared with an annual increase estimated for the 1990s at 1.5 to 2 per cent for dairy farming and probably less than 1 per cent per annum for the meat sectors without their use. In the United States it has been estimated by the Office of Technology Assessment[14] that wheat productivity would in the period to 2000 increase by 1.5 per cent per annum without biotechnological innovation and 2 per cent with. The proportionate increase would be similar, or less, with the other main crops.

In the European Community a survey by the Bureau Européen des Recherches in 1989[15] suggested an additional 0.5 per cent a year productivity increase for the application of biotechnology to the cereals sector; the effect for the application to the livestock sector was obscured by the expectation of substantial political obstacles being erected against the full exploitation of new techniques – particularly in dairy farming.

Box 9.1 OECD view of specific effects of biotechnology on agriculture

1. Substitution of old methods of input production by new biotechnologies – for example, tissue culture in plant-breeding to produce homozygous plants – could substantially reduce breeding time and costs.
2. New substitutes for farm inputs, such as insect- or pest-resistant plants to replace insecticides, could increase competition and therefore reduce input prices.
3. New biotechnology firms entering the input supply market could also increase competition.
4. New markets for agricultural products could be created, particularly in the field of non-food and renewable resources, but also for food and feed products.
5. Alternative uses for existing products could be developed, such as bio-ethanol for grain and other arable crops.
6. New competition for existing products by substitutes such as single-cell protein is likely to emerge.

However, it is not only the production of food that could benefit from the application of biotechnology: food-processing could also gain substantially from the new science. The food industry is, in any case, the oldest user of biotechnological techniques – brewing, wine-making and bread-making spring to mind. Biotechnology in the modern sense, however, offers the food industry a wide range of new techniques that could be used to cut costs, raise productivity, increase consumer choice and improve food safety. The OECD[16] identifies a number of factors that will decide how far these new technologies and techniques are adopted. A major use is to provide substitutes for existing foods. The production of isoglucose and other 'sugar substitutes' from starch is an example of one already successful commercial development. It will increasingly be used to provide substitutes for rare food ingredients difficult to obtain, such as certain herbs and spices.

Contrary to the generally perceived view of modern scientific development in the food industry, it is probable that biotechnology may be able to be used to reduce the levels of added chemicals and the level of processing. It can be used to improve the nutritional value of food, too. New methods of detecting contamination and quality deterioration will become increasingly important.

It will, however, be increasingly difficult for people to accept the application of these new biological techniques to food production. Such techniques as enzyme processing and genetic engineering arouse considerable public concern, and it is this type of potential opposition which the regulatory system will have to overcome.

> Biotechnology in food products, ingredients or processing will raise new regulatory issues. Well known ones would include the use of genetically modified food·bacteria, but others would arise when food components are made in plant cell culture or microalgal culture rather than in normal crop production, while in food processing, the development of enzyme conversion systems may also raise new regulatory questions. These questions will encompass basic confirmation of safety and acceptability and also specific requirements concerning labelling and designation of products. Regulators will need to balance the concerns about new technology against the benefits that may arise from biotechnology in relation to reduction in chemical additives or the improvement of preservation of food products.
>
> (Biotechnology, Agriculture and Food, OECD), Paris, 1992

Recent advances in microbiology, genetics and enzymology are expected to remove current yield and purity barriers to the use of biotech methods. The production of new types of fermenters and control systems means that bioprocess engineering will increasingly overcome the physical exploitation of the new techniques. Recent advances in enzyme technology include the development of enzyme bioreactors, enzyme encapsulation and new enzymes made by protein engineering. Protein engineering will allow the alteration of food enzymes at the molecular level.

One of the most promising areas for the application of biotechnology to the food industry is the preservation of food: an example is the genetic engineering of lactic bacteria with anti-microbial features. Biotechnology can also be used to convert low-value agricultural products into high-value foods. The production of myco-protein through fungal activity is an outstanding example.

The application of biotechnology to agriculture – if not obstructed by politicians – should lead to downward pressure on food prices and speed up structural change in agriculture. It is also likely to increase vertical integration in the agriculture and food sectors. The use of these technologies may also add to the existing bias in competitive advantage in agriculture and food to those countries with large farms and high scientific and managerial competence. The OECD (1992) concludes:

> Biotechnologies are related or relevant to nearly all the major agricultural policy questions of our time . . . [and] will increasingly become central to technical change in agriculture and food production, agricultural policies will have to cope with [them] on a permanent basis by following, selecting and using them.

Opponents of the application of biotechnology to agriculture would do well to reflect upon the warning from agricultural scientists consulted by the OECD. They are convinced that biotechnology will be essential to the maintenance of an adequate food supply for a world population which will continue to grow for some considerable time.

10

EUROPEAN AGRICULTURE IN THE TWENTY-FIRST CENTURY

The European Union will undoubtedly become a much larger organisation during the rest of the 1990s and in the following decade. The reason for this is relatively obvious: the political pressures on the Community for reasons wider in scope than agriculture or agricultural trade will force it to allow other countries to join the Union. The reasons for this enlargement are, to a great extent, to do with the future stability and security of Europe. It is regarded as impossible for the rich, western section of the European continent to ignore those countries, relatively prosperous, still remaining outside the Community in the western part of the continent, but also, and probably more importantly, to ignore those countries in central and eastern Europe which, since the collapse of Soviet hegemony, are increasingly seeking a place in the new European order.

The so-called 'Visegrad Four', namely Poland, Hungary and the two republics which made up the former Czechoslovakia – the Czech Republic and Slovakia – are countries which historically have more in common with western Europe than with their eastern neighbours and still see themselves in this light, despite the forty years of tyranny which since the end of World War II has blighted their industrial production and physical environment.

The majority of present EU member countries regard the absorption of the central European countries as essential to the future stability of the whole of Europe. The eventual incorporation of the more eastern countries and of the republics of the former Soviet Union (FSU) itself is regarded as being a long-term goal too. This is considered to be something which is less essential. There is still significant scope for the eventual formation of some eastern European economic union centred on the Russian Republic, the Ukraine, and several others of the republics of the FSU.

The absorption of the central European countries of Poland, Hungary, the Czech and Slovak Republics – the so-called Visegrad Four – is regarded as being something which should be achieved as soon as possible: it is, however, unlikely before the first decade of the next century, since the European Union is itself undergoing considerable economic and political reorganisation following the unification of Germany and the problems of the creation of an internal integrated economic and monetary system. In the meantime, however, the EU12 has committed itself to the absorption of most of the remainder of the European Free Trade Area left after the most recent enlargement of the European Union in the mid-1980s, when Spain and Portugal followed Greece into the European Community.

Finland, Sweden and Austria became members of the European Union in January 1995, leaving only Switzerland, Iceland and Norway as remaining members of what once was the seven-country European Free Trade Area. The addition of the EFTA Three to the EU12, and thus the creation of the largest developed economic grouping in the world, will not, however, make any significant differences to the CAP or to agriculture and the food trade within the enlarged European Union of fifteen. This is essentially because the new member states all have very similar agriculture structures – and similar problems. It is also because their agricultural trade is to some extent already integrated with the Community but, more importantly, because their food and agriculture markets are in a state of relative equilibrium. In other words, they in general consume as much food as they produce. This applies to most of the major commodities, with the possible exception of sugar in which they are in deficit and also of wheat, where Sweden, exceptionally, is a significant exporter.

There is also a highly developed trade in processed agricultural products between the new member states and the existing Union. This has been a significant feature of the closer grouping of trade between the EFTA countries and the European Union in the short-lived European Economic Area, established in the early 1990s as a half-way stage to full incorporation of EFTA into European Union.

The EFTA Three countries also had already begun to modify their agriculture policies for two reasons: first, to meet the inevitable adjustment of their agricultural market support systems and trading arrangements in order to meet the expected GATT agreement on agricultural trade; and second, since these countries were also determined to join the European Community, they needed to align their policies with those already existing in the EU12. For Sweden and Austria this adjustment was very much less of a problem than for Finland. Both countries had begun to reduce producer supports and to change the basis of their agriculture policies so as to take the emphasis off market intervention, and where necessary to use the type of direct subsidisation which became more popular within the European Union as a result of the 1992 CAP reforms. Sweden, for example, had already largely modified its policy so as to be able to fit into the European system without any major change in farmers' incomes or the way in which its agricultural trade was conducted.

For Finland, however, there are substantial problems because it has large numbers of landholders in the Far Arctic and Near Arctic north, therefore needing substantial direct social subsidisation. The level of subsidy which these areas and the farmers in them receive exceeds even the substantial subsidies paid in the less-favoured areas of the European Community. The maintenance of this highly socialised agricultural/rural policy proved to be a major source of contention between the applicant members and the European Union during the negotiations for membership. It is, however, quite clear that the European Union of twelve is not only prepared to accept the methods and level of subsidisation of the remote regions of these Nordic countries, but is also probably likely to adopt similar methods for dealing with its own social rural problems, particularly in the less-favoured Mediterranean areas of the Community in Italy, Spain and Portugal.

It is therefore clear that, as with previous enlargements of the Community, the European political and economic agglomeration changes not only in its geographical and political form, but also in the operation of any policies which may be operating at the time when new members join the community. Thus, as the Union enlarges, so its policies – and in particular, agriculture and rural policies – change to suit new conditions. The addition of Scandinavian countries affects agriculture and food production in other ways as well, particularly in the areas of environment, animal welfare and food health and safety standards.

The standards for handling livestock, the transport of livestock, the operation of slaughterhouses and the rules for the safe handling of food are in general higher in the Nordic countries and in Austria than they are in many of the existing members of the European Union. The lower standards within EU12 for the control of the spread of salmonella from livestock to humans was, for example, a major problem for Sweden during the 1994 membership negotiations. It is therefore likely that in the years to come the EFTA Three will combine with Denmark, Germany and, possibly, Belgium, in demanding higher European standards not only for the agricultural environment but also in the general standards of food trading and animal welfare.

Enlargement of the European Union to include the former three EFTA countries is unlikely to make any significant difference to the balance of the EU's agricultural markets. Overall, the three countries have more or less a balance between production and consumption of the major agricultural products.

The addition of the Visegrad Four central European countries to the EU16 will be very different from the 1995 enlargement. There are several important reasons for this: most importantly, the problems of reorganisation of these countries after the economic collapse following the fall of communist government systems, the serious agricultural and environmental problems which exist in several of the countries, most notably in the Czech and Slovak Republics, and lastly and probably most important in the medium to long run, the significant potential for the increase of production in these countries.

All four of these countries have important agriculture industries which are capable of expanding production, and expanding in a very short time. What is lacking at present is not only capital investment to repair the damage done by the communist regimes, but most importantly, the demand for agricultural products to stimulate that investment and to encourage farmers to increase production. The land in these countries is fertile and can easily be brought into a state of high fertility; there is to a great extent still a significant work-force sufficiently educated and trained to be able rapidly to apply new methods and techniques of crop production and animal husbandry.

Agriculture is vitally important to the economies of central and eastern Europe. It represents a substantial proportion of the value of economic activity, it employs large proportions of the working population and, in most countries, it provides an important proportion of export revenues. For several of the central and east European countries expansion of agricultural exports could provide the most rapid method of increasing foreign currency earnings and raising national income. Most of the countries have substantial agricultural resources which are currently massively under-exploited. Many obstacles remain, however, to the exploitation of this productive and trading potential.[1]

The most important short-term problem is the depression of demand for agricultural products within the domestic economies, caused by the removal of the consumer subsidy and state buying operations of the former communist regimes. Without the stimulus of expanding domestic demand, agricultural production has generally slumped in most of the CEECs.

In several of the countries, too, the process of economic reorganisation has probably affected the agriculture and food sector even more than industry. Rationalisation of the process of transferring the ownership of land from the state to private individuals and putative commercial entities has proved every bit as difficult as expected when the communist command economies collapsed.

The central problem which the governments of the CEECs now face is the reconciling of the need to protect and nurture the agriculture sector into a fully productive and competitive state, while at the same time avoiding falling into the trap of establishing interventionist policies similar to those of the previous regimes or even of the type from which the European Union is trying to extricate itself in the later 1990s.

Undoubtedly, agriculture in central and eastern Europe in the early 1990s was in a parlous state: production was falling, marketing and distribution was disorganised and farm product export revenues were declining. By the mid-1990s however, there were signs of recovery.[2]

A serious problem for the EU15 is that the cost of operation of the EU's CAP could very nearly double in a Union enlarged to take in not only the EFTA countries, but also the Baltic states and the central European countries in the Visegrad group.[3] More conservative estimates[4] suggest that the proportionate increase could be smaller. But certainly expansion of production and the decline in consumption in these ex-communist potential members could lead to a steep rise in the cost of disposal of surplus production. The heavy

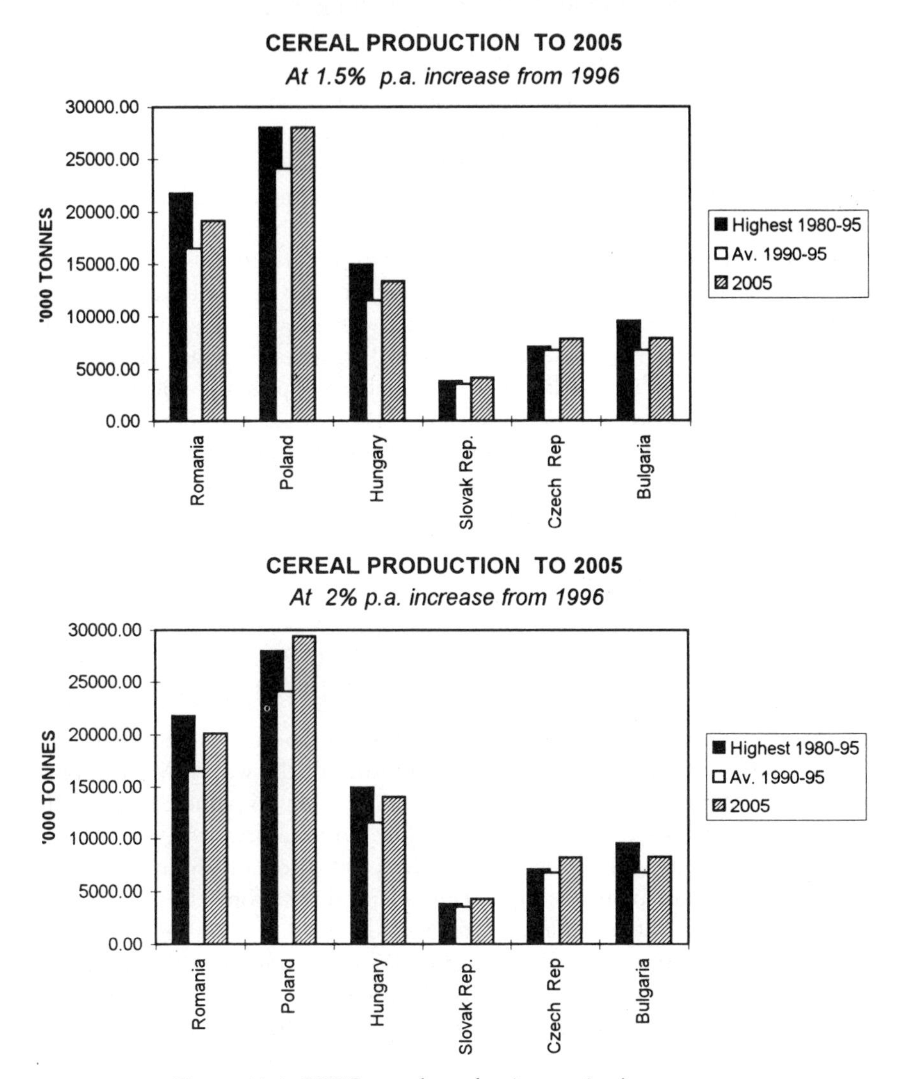

Figure 10.1 CEEC cereal production projections

dependence of these countries on the mainstream northern products means that they are likely to be major producers of cereals, meat and dairy products already heavily in surplus.[5]

Although the proper and full application of the 1992 CAP reforms will moderate the impact of enlargement on the EAGGF Guarantee budget, expected savings on export subsidies as a result of the reforms could, however, be counterbalanced by large farmer compensation payments. In the absence of cuts in export subsidies, the cost of dumping EU surpluses on world markets could rise from current 19 billion ECU to 55 to 60 BECU a year in a Union of eighteen countries[6] according to what is generally regarded as one of the more 'far-out' estimates of the CEEC membership impact. This estimate is based on a Community enlarged to take in the EFTA countries, the Baltic states, Poland, the Czech and Slovak Republics and Hungary. It is, however, unlikely that the Union will contemplate the addition of the Baltic states until the end of the first decade of the next century.

The essence of the problem of reconciling the agricultural systems of central and eastern Europe to that of the EU is that the CEECs have large areas of fertile land and livestock potential which can produce a great deal more of the commodities which currently receive the highest support within the EU. What is clear is that the EU will not be able to apply the policies operating in the EU15 in the mid-1990s to the EU19 or EU21 without serious danger of provoking increased production and budgetary burdens. The solution to the problem is seen by some commentators to lie in the establishment of an internal integrated agricultural market between the European Union and the ex-communist countries of central and eastern Europe – in advance of any full addition of the CEECs to the EU.[7]

THE EU AND THE LESS DEVELOPED COUNTRIES' AGRICULTURE POLICIES

The European Union has significant and important agricultural trading relationships with the less developed countries of the world. Most important are the arrangements under the Lomé Convention with the countries of the African, Caribbean and Pacific (ACP). The Community, despite its protective agriculture policy which encourages the production of almost every major commodity, still manages to squeeze in products from the less developed countries, often at preferential rates and designed to give these countries a relatively wealthy market in which to sell their products. Although the bulk of the agricultural exports from the ACP countries are of tropical products which cannot be grown in the European Union, the Community does make one major concession which is important to countries in the British Commonwealth and the French Overseas Dominions: the annual Lomé quota for the import of 1.3 million tonnes of sugar.

This sugar quota provides a major boost to the incomes of producer economies in the Caribbean particularly, but also in French dominions, such as Mauritius. The important feature of the concession is that the ACP countries are guaranteed the same price for their cane sugar as that received by European beet-sugar producers. This means that for an important part of their output, they are guaranteed a price which is on average at least 100 per cent greater than the price which can be obtained on the residual world market. The Community also operates a preferential tariff system for the import of bananas from the less developed countries. This is organised so as to give preference to countries in the ACP area and to a considerable extent discriminate against developing countries of Central and Latin America which do not have the same preferential access to the European market.

EU AGRICULTURE IN A NEW WORLD TRADE ORDER

Clearly, European agriculture policy has to undergo considerable further modification to fit into the 'new world order' created by the conclusion of the first real multilateral agreement on agricultural trade, by the collapse of the communist hegemony in eastern Europe and by the growing awareness of the need to involve the less developed countries in the world agricultural trading system. Even without the outside pressures, the EU's common agricultural policy has to adapt to a changed configuration of alliances and relationships within Europe; it also has to adapt to the changed demands and attitudes of its own people. The dilemma has been well stated by a group of European economists hired by the EU Commission in 1993-94 to advise it on the future path of development which EU agriculture policy should take in the later 1990s:[8]

> Fundamental technical and economic forces will continue to pose inescapable challenges to which the political process will have to adjust. Although the lower EU prices now being introduced may have a longer term effect of slowing down the pace of invention, development and adoption of new methods, technological progress such as genetic manipulation and electronic controls will continue to raise the productivity of even the most advanced production systems. There is also still considerable scope for productivity gains through structural change (larger farms). In addition, hitherto less efficient regions of the world such as eastern Europe have a catch-up potential which will need to be accommodated. Substantial supply increases in agricultural products worldwide must therefore be expected. On the other hand, demand for agricultural products will remain stagnant in the Community while increases in effective world wide demand will be limited.

Thus, while technological advance will have the potential to continue to push up output – whatever policies are operated – production could continue to rise. Policies based on rewarding farmers for output – which even the 'reformed CAP' continues to do, can do nothing but increase output which cannot be absorbed by the market. Continuation of such a policy can only create further budgetary, environmental and social problems. As the EU expands its number of member countries and therefore its farming area, so it has to guard against offering new incentives to expansion to enterprising farmers in the ex-communist countries.

EUROPEAN AGRICULTURE AND THE WORLD FOOD SYSTEM

The major, continuing question[9] for the EU agriculture industry and for European policy-makers is: what is the place of European agriculture in what will increasingly become a more integrated world food production and supply system during the coming decades?

European agriculture undoubtedly faces a number of challenges during the next ten years: the later 1990s and the first half of the first decade of the twenty-first century. The most important is likely to be to adapt its production to new, lower price levels, while at the same time maintaining economic and social viability in an environmentally acceptable manner.

In addition, the agriculture of a European Union, enlarged to include at least the four main agricultural countries of central Europe, will have to adapt to a new role in world agricultural trade. A constant preoccupation of the policy formulation process is likely to be the Union's role in the world food complex. A major preoccupation is likely to be: what is the world's demand for food likely to be in the first half of the twenty-first century and how much of this demand could or *should* be met from EU production? And to fulfil the decided role, what will be the price to the taxpayer, consumer and the environment?

Central to these challenges for the EU is the question of the future productive capacity of world agriculture in relation to expected population growth. If the ratio of food supply to population is in fact declining, as some analysts maintain, then it may be necessary for the EU to abandon all controls and limitations on agricultural production and to expand production at any price. At any price?

The question of relative and comparative costs of production in Europe and other parts of the world is a consideration which has been nearly at the bottom of the list of the criteria for EU agriculture and food policy formulation – if not actually off the list entirely. Certainly, the economic cost has only rarely been considered; the environmental cost has been considered, and then only recently, in a very coincidental manner to the main preoccupation of sustaining production and maintaining the incomes of agricultural producers. The central, global question is, however, the production capacity and sustainability of capacity in the world agricultural industry in a period of world population that is still rising.

In 1972 the Meadows husband-and-wife team published their now famous work, *The Limits to Growth*.[10] A great deal of this 'Club of Rome Report', as it was more generally known, was devoted to the question of future food supplies: it predicted that there was a serious danger of the world being incapable of producing enough food to feed its inevitably increasing population. The food crisis would come about as early as the first decade of the next century. The main thrust of the Club of Rome argument was that

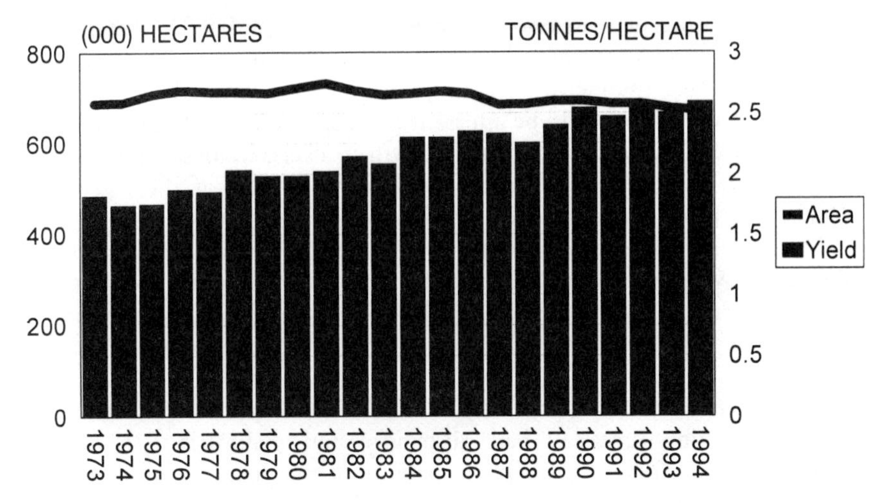

Figure 10.2 World cereal area and yield, 1973–94

Source: USDA PS and D database

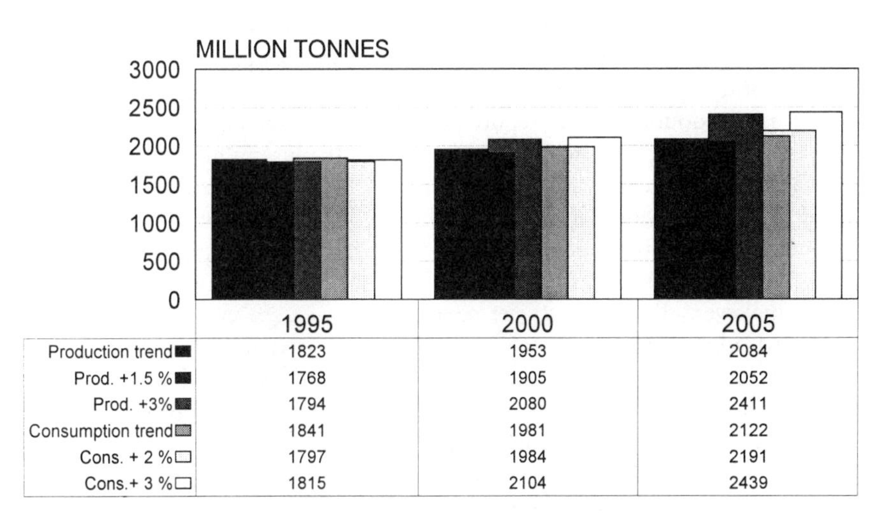

	1995	2000	2005
Production trend	1823	1953	2084
Prod. +1.5 %	1768	1905	2052
Prod. +3%	1794	2080	2411
Consumption trend	1841	1981	2122
Cons. + 2 %	1797	1984	2191
Cons.+ 3 %	1815	2104	2439

Figure 10.3 World grain production projections to 2005

the expected increase in population could only be fed by further intensification of the already intensive methods of crop and livestock husbandry being applied in America and Europe at that time. The 'limit', however, would be that such methods could not be sustained. The types of cropping systems used were rapidly exhausting the soil's fertility in most developed countries, and the concentrated uses of fertilisers and crop protection chemicals were accumulating a level of residues which were unacceptable to the healthy survival not only of man but also of wild animals and flora. Inevitably, it was argued, limits on these 'pollutants' would prevent food supply from being expanded to feed the increase in hungry mouths. It was predicted that such limits on production would be apparent as early as the 1990s and that there would be a tendency for population growth to overtake food supply growth by the mid-1990s. Despite one or two worrying years at the end of the 1980s, this prophecy has proved largely unfounded.

More recently, the Meadows team published a follow-up to *The Limits to Growth.*[11] While extending the timetable for possible catastrophe, this later work was a much more measured and less alarmist analysis of the constraints upon food production by what are regarded as 'conventional' intensive methods.

This later analysis suggests that the world's hungry are the victims of wasteful farming practices, bad husbandry and maldistribution of food supplies – rather than any incipient inability of the planet to produce food. A great deal of the report is devoted to the vital relationship between people and their food supply: 'The amount of food grown in an average year is sufficient to feed the present world population adequately, but not lavishly. Because of waste and unequal distribution, it feeds part of the population lavishly, part moderately, and another part totally inadequately.' The ability of the world's agriculture to feed the world and even to match population growth is not doubted. The Report points out that between 1950 and 1985 world grain production trebled to around 1800 plus million tonnes a year – an annual growth rate of 2.7 per cent, and faster than the rate of increase in the number of people in the world. In 1989, had the food produced been adequately distributed, it would have fed 5.9 billion people a subsistence diet, 3.9 billion a moderate diet or 2.9 billion at the European level. This figure, however, allows for 40 per cent wastage.

The Meadows team omit to point out – but it is none the less mathematically obvious – that if wastage were reduced to European levels, available food supplies would increase by 35 per cent and almost everyone could enjoy a European standard of nutrition. World availability of food would increase well beyond current levels if production in the less developed countries were moved closer to European yields. As it is: 'Of the earth's more than 5 billion people over 1 billion at any one time are eating less food than their bodies require' (see note 11).

As a recent FAO study of 117 countries of Latin America, Africa and Asia showed,[12] only nineteen of these countries (with less than 2 per cent of the world's population) would not be capable of feeding their year 2000 population if all their technically available land were fully exploited. The computer model (World 3/91) operated by the Meadows team indicates that even if average grain yields remain at current levels, the amount of land available for cultivation will exceed the amount needed to maintain adequate food supplies well beyond 2100. More optimistic agronomists – but not necessarily the Meadows team – would argue that average yields will tend to increase, whatever limits may be imposed on fertiliser and pesticide use.

If they are right, then the margin of available land is likely to be more than adequate. The follow-up Meadows Report is, as its title suggests, about the dangers of the expansion of population and the human demand for food and other resources beyond the limits of feasible supply. Much of the land at the 'margin' would be extremely costly and difficult to bring into cultivation – particularly if environmental costs are brought into the equation.

The important question that has to be answered therefore is: will human population and therefore demand for food exceed the capacity of the world's farmland to produce on a sustainable basis? An important supplementary question is: will agricultural practices in some parts of the world continue to be so wasteful that the amount of land is reduced by the creation of wastes?

In 1650 there were probably 0.5 billion people in the world and the population was increasing at the rate of 0.3 per cent a year. Had this rate of growth been maintained, the world population would have doubled in 250 years.[13] By 1900, however, the population had reached 1.5 billion and was increasing at 0.5 per cent a year. By 1970, the rate of increase had reached what the Meadows team describe as 'super-exponential'. Fortunately, in the following two decades the rate of growth itself did fall; none the less between 1971 and 1991 population increased from 3.6 to 5.4 billion.

Growth is still exponential, and it is this phenomenon which is seen by the Meadows team as the threat to the planet. The world's point of inability to sustain the human population is seen as 'probably not much more than a doubling [of population] away'. The mathematics of population growth suggest that on the present trend rates, this could happen by 2025 – if no checks to population growth are introduced and the growth rate remains at the current approximate 2 per cent a year.

Inevitably, what might be regarded as the 'Club of Rome school' has always been under intense attack from the optimists. The undoubted leaders in this field were Herman Kahn and his Hudson Institute group of analysts. Their basic conclusion was that the world's agricultural resources are more than adequate to feed ten times the current world population and that the increased use of fertilisers and crop chemicals needed to sustain higher rates of production do not represent a serious threat to the habitats of either man or animals. More recently, the heirs of Herman Kahn concluded at a 1991 conference on the state of agriculture that there is no problem in maintaining the world's food supply that cannot be solved by modern agri-business methods.[14]

In a paper to a Brussels conference in the summer of 1994 the Director of the Hudson Institute's Global Food Studies Centre Denis Avery,[15] took the controversy into the environmentalist's camp, arguing that only modern intensive agriculture would guarantee adequate food supplies and protect the environment. Extensive, 'organic' methods could be expected to increase pressure on the natural environment – principally by increasing soil erosion – and would fail to maintain food production at the level necessary to keep up with population increases.

Avery reasons that while the world's farmers are actually cropping approximately 5.8 million square miles of land (equal to the land area of South America), they would have to cultivate almost three times that amount to cope with the increases in food demand since the 1950s – if it were not for modern farming methods involving high levels of artificial fertiliser and pesticide use. Avery argues:

> Chemically-supported, high-yield farming is already saving roughly 10 million square miles of wildlife habitat from being ploughed down for food production. . . . The true long term threat to human existence is soil erosion, and the new farming systems are cutting soil erosion by 50 to 98 per cent.

If the world population did reach 10 billion, the world-wide abandonment of modern farming methods and the adoption, instead, of extensive, low-yield farming could need as much as 25–30 million square miles of crops to feed it. This would be the equivalent of the entire land area of North and South America, Europe and most of Asia put together. Clearly, this is an impossible option; the solution must lie in farming the land already under cultivation more intensively.

> High yield agriculture has been raising world crop yields by 2 per cent annually for the past 30 years. It is currently raising food production in the Third World at roughly twice the population growth rate in those countries. Nor is there any slackening in productivity growth. If the world continues to fund research in high yield agriculture – especially with biotechnology – we should be able to feed the 21st century human population with less crop land than is ploughed today.[16]

Not surprisingly, since it is politically correct only to believe that modern agricultural science is essentially malign, what are popularly judged to be these 'extreme optimist' arguments have been even more under attack than those of the Doomsday school. It is probable that the realistic picture of the world's likely food production resources lies somewhere between these two extremes.

Some five years after the publication of the *Limits to Growth* Report, a group of Dutch economists, agronomists and policy analysts concluded that all the world's population could be easily fed through the more efficient utilisation of the world's existing agricultural resources.[17] In other words, if the land already in production were cultivated more efficiently, there need be no shortage of adequate supplies of carbohydrate and protein – even for a world population increased by 50 per cent plus on the mid-1970s level of 5 billion.

That increased research into more productive and less destructive methods of agriculture is, however, undoubtedly needed to sustain food supplies of an increasing world population was stressed in the World Bank's 1992 *Development Report*,[18] which put special emphasis on 'development and the environment'. The generation of new knowledge, it said, is the 'most potent and least costly avenue to improving productivity'.

Increased food production to meet the needs of an inevitably rising world population is likely to come from raising output per hectare, rather than from any significant increase in farmed area. The Bank points out that extension of the world's farmed land area can only lead to greater environmental pressure on forests, wetlands and other vulnerable and vital areas of the planet. In contradiction of the popular views of the international green lobby, the World Bank indicates quite clearly that developing countries will have to go the same way as the developed countries in intensifying crop rotation and other husbandry methods. Both developed and under-developed countries must, however, abandon the environment-threatening practices of high fertiliser use and indiscriminate pesticide application.

This view of the future path of the source of increased world food supplies – from increased production on existing land rather than from any significant extension of the farmed land area – is the one area where the 'neo-Malthusians' and the optimists of the Hudson Institute and others agree.

The World Bank stresses that food supplies can, however, only be sustained and increased by maintaining intensive agriculture:

> Ninety per cent of the doubling in food production over the past quarter century came from higher yields and only 10 per cent from cultivating more land. Intensification, which will account for most future increases in production, will create environmental problems. The right policies are of two types: those that enable farmers to do what is in their own interests, such as managing soils better, and those that provide incentives to stop behaviour which primarily hurts others.

The World Bank argues that although world population may grow by more than 50 per cent to 9 billion over the next forty years, it is quite possible for food production to expand to meet fully the increased demand – and without environmental degradation:

> To match this increase, world grain output will have to grow by about 1.6 per cent a year – a difficult target, but less than the 2 per cent a year increase achieved over the past three decades.

In the future, yield increases will have to be sustained by improved genetic material in crop plants, combined with more 'biological' pest control. There will have to be a trade-off between increased productivity and some modest expansion of farmed area in some parts of the world. In 1990 1.9 billion tonnes of cereals were produced on approximately 700 million hectares of land in the world at an average yield of around 3 tonnes/hectare. In the year 2030, 3.6 billion tonnes of grain will be needed. Theoretically, the production combination choice will be between cultivating 1.4 billion hectares at an average yield of just under 3 tonnes/ha at one extreme, or cultivating 500 million hectares at an average of around 7.5 tonnes/ha at the other. The World Bank concludes:

> Meeting the doubled food demand that is anticipated by 2030 will be feasible but will require substantial productivity gains. Fundamental to meeting the challenge of increasing productivity will be better application of existing, but under-used, knowledge about resource management and development of new agricultural technologies.

The pace at which new environmental legislation is imposed on world and European agriculture will, of course, have an important influence on the extent to which food supplies will or will not match population growth.

Great though the strides in 'biotech', non-chemical pest control may have been in recent years, the big breakthroughs in genetic and bug-eat-bug plant protection will not come to rescue farmers in a chemical-free world from insect, disease and weed depredations. The largely conventional use of herbicides, insecticides and fungicides will be needed to maintain and increase food supplies for a decade or two yet to come.

The world now faces a situation, however, where the world population is expected to expand from current 6 plus billion to 8.5 billion by the 2020s. Will the world's agriculture be able to meet this demand in a sustained and sustainable manner?

What is certain is that what might be regarded as the 'Total Organic Solution' dreamed of by the extreme 'greens' would lead to an increasingly serious food shortage, since it would diminish the export surplus of efficient world agricultural producers such as the United States, Canada, Argentina and Australia, which currently provide the buffer stocks in world food supplies that are the main insurance against much more widespread famine than already exists in undoubted food supply crisis areas.

A leading independent economic consultancy in the United States, GRC Economics,[19] has analysed the effects of 'US Agriculture without pesticides or artificial fertilisers'. The result, it estimates, would be a 45 per-cent increase in the cost of food and a 50 per-cent decrease in the supply of fruit and vegetables. The elimination of pesticides alone would result in a 20 per-cent reduction in US wheat production, a 25 per-cent drop in maize (corn) production and a 27 per-cent fall in soyabean output. In other words, that part of US farm output that forms the 'safety buffer' in world supplies would disappear.

It is of course unlikely that the United States or Europe – or any other developed country or economic grouping – will ever go as far as banning the use of chemicals completely. What the agrochemical industry fears most, however, from the present green and consumerist pressures is that the regulatory system will become even more ponderous and long-winded in its processes than it is already.

The demands of modern agriculture – in less developed countries as much as in Europe and North America – put increasing pressures on scientists to produce wider ranges of better products. Consumers and environmentalists also demand 'better' farm chemicals, products that do not contaminate food or the habitats of people and animals. Inevitably, these objectives, though they need not be irreconcilable, lead to conflict. In the United States it now takes the best part of ten years and more than $40 million to get a new chemical through the vetting system and onto the market; in Europe the costs are even greater. When the EU Council of Ministers finally approves a planned EU Pesticides Directive, it is estimated that more than 60 per cent of the currently listed crop chemicals will disappear from the market.[20] This is because the costs of re-testing and re-registering will be greater than any profits that could be made from keeping them on the market. Many environmentalists would argue that this is a very good thing, since many of the chemicals which will disappear are the 'dirtiest' and most inexact materials which pose the greatest environmental hazards.

There is too the more recent bogy of global warming. Modern developments in European agriculture, principally the reduction in the numbers of trees and the increase in livestock numbers, have made their own contribution to global warming through, respectively, the reduction in the capacity of the Earth's surface to absorb carbon dioxide and the increase in the methane discharge into the atmosphere. It is expected that, assuming no significant change in the level of emission of anthropogenic 'greenhouse gases' (commonly known as the 'business-as-usual scenario'), there will be a doubling of the carbon dioxide concentration in the atmosphere by the mid-twenty-first century. This is estimated to lead to an increase in average European temperatures of 0.5 degrees centigrade by 2000–2010, an increase of 1.5 to 2.5 degrees in the following three decades and a further 3 degrees in the following half-century.[21]

There are likely to be two main effects on agriculture: (1) a fertilising effect resulting from the greater carbon dioxide concentration in the air; and (2) a weather effect on crops and livestock. Carbon dioxide enhancement is likely to lead to yield increases in the major food crops – wheat, barley, rice and potatoes. It should be borne in mind, however, that the benefit of this stimulatory effect will only be obtained by higher fertiliser use.

The most important effect for Europe will, however, be an extension of the growing season and higher average temperatures in northern Europe; at the same time, a shorter growing season could be expected, but with higher temperatures and less rainfall, in Mediterranean areas.[22] Such a change would lead to a northward shift of crops traditionally regarded as products of the southern areas. A 1.5-degree rise in average temperature would shift the area where, for example, sunflowers could be grown 500 kilometres northwards and maize would be able to be grown for grain across most of western Europe.

A reduction in agricultural capacity in traditional food-exporting countries could, however, put European agriculture under increased pressure to produce more – thus reversing the present trend away from increased intensive production and towards more environment-friendly developments.

A less optimistic view is held by UK climatologist Martin Parry, who sees global warming leading to a substantial increase in famine, due to reduced crop yields in Africa, tropical Latin America, India and South-east Asia. Parry believes[23] that most of the world's food crops of importance to the less developed countries – rice, wheat, sorghum and millet – will be seriously affected by the expected change in climate. Parry, Director of the Environmental Change Unit at the University of Oxford, has made his assessments on the basis of the British Meteorological Office and the NASA Goddard Institute of Space Studies computer models. He argues that global warming (on the basis of an assumed doubling of carbon dioxide content in the atmosphere) will result in declining crop yields not only in central latitudes, but also in North America. Such reductions in North American yields are expected to have an important effect on world food supplies.

Parry's study, like the USDA analysis, incorporates assessment of the adjustability of world agriculture to climatic change. The main findings of his assessment are: (1) declining harvests in all parts of the world except the most northern latitudes; (2) a rise of 400 million in the number of people at risk from hunger; (3) an increased demand for irrigation water in Europe; and (4) the destruction of farmland by the rise in the sea level and by salination of fertile coastal areas.

The threat of reduced food supplies on international markets might be used as a justification for higher EU production, bringing with it increased fertiliser and pesticide use. Higher temperatures would mean increases in fungal diseases of plants. Many of the major weed species will grow faster and more abundantly. These developments will therefore increase the pressure for heavier use of fungicides and herbicides.

What is unlikely is that, though there are substantial obstacles to the sustainable increase in world food production which will be needed to feed an increased population, the 'inadequate carrying capacity' prophecies of the food supply pessimists will be proved to be true. There are very plausible counter-arguments to all of the main elements of their case.

What could be described as the merely technical analyses of the world's carrying capacity, such as that carried out by the World Resources Institute (WRI)[24] and by Lester Brown and his team at the Washington Worldwatch Institute[25] are challenged by analyses which emphasise the political as well as the technical factors influencing increased food production. Gerhard K. Heilig of the International Institute for Applied Systems Analysis challenges[26] the basic assumption that the world's carrying capacity is limited: 'The carrying capacity of the earth is not a natural constant – it is a dynamic equilibrium, essentially determined by human action.'

Technology and man's ingenuity should allow food production to expand to meet increased population. The most important element of this counter-argument is that the world is not short of fertile soil to increase production: most constraints are specific natural conditions which can be overcome with modern agricultural technology. Heilig fundamentally challenges the World Resources Institute assessment of world food production capacity:

> WRI data on soil constraints are worthless as indicators of the earth's carrying capacity: they do not match with current trends in food production – to be more precise in some cases the indicators are just absurd when compared with agricultural performance.

For example, the WRI claims that India has only 33.2 million hectares of unconstrained soil and yet 169 million hectares of cropland is currently being cultivated. What is more important, there is no evidence of marginality of this additional area – India's cereal production increased 129 per cent between 1961 and 1989 – with cereal yields more than doubling to 1.92 tonnes/ha. Countries with so-called 'unconstrained soils' (according to the WRI assessment), such as Chad, do not have high output – suggesting that the failure to deliver is human rather than finite. Heilig says there is simply no correlation between food production and soil constraints as reported by WRI.

Heilig argues that there are numerous ways of increasing production on existing arable land: (1) expansion of the area of multiple harvests; (2) increased production on marginal land; (3) the extension of food production to the water – lakes, rivers and seas; and (4) switching from non-food to food crops.

> It is the level of agricultural technology which determines the land's food production capacity. This basic understanding is still rare among today's environmental doomsdayers, such as the WRI. They continue to focus their attention on the physical condition of soils, collecting ever more detailed inventories of soils characteristics. But they are obviously blind to the fact that these characteristics are becoming irrelevant. The size and quality of soils are just two variables in a multi-term equation of agricultural productivity, which is mainly determined by technological, economic, socio-cultural and political factors.

The International Institute for Applied Systems Analysis (IIASA) does not believe that shortage of water is a serious limitation on agriculture, and that it is certainly not a serious limitation on the globe's carrying capacity. The undoubted desertification problems in Africa are amplified by unsustainable practices.

Energy consumption in agriculture is unlikely to be a limitation on increasing production, for the simple reason that 'modern agriculture does not consume large amounts of commercial energy'.

The notion of physical limits to growth is a faulty concept:

> If we take into account the creative potential of man, there is no foreseeable limitation to the basic natural resources of food production, which are space, water, climate conditions, solar energy and man-made inputs. All these resources are either unlimited or can be expanded, better utilised or re-designed to a very large extent.

In some developing countries the problem of soil degradation and reduced production is the result of too little fertiliser use rather than too much. It is colossal policy failure which has caused widespread under-nutrition and famine during the past four decades. 'The scandal of famines in Africa is not a result of agriculture approaching carrying capacity; it is mostly a consequence of massive policy failures, corruption, ethnic conflicts, ignorance and incompetence of ruling elites.'

The evidence, therefore, suggests that farmers will be able to go on producing more food faster than the mouths to eat it can increase for the foreseeable future. The big problem of the world's food system is, however, not production, but the maldistribution of more than adequate production. As too many television documentaries have testified: the rich have too much food and the poor have too little. It is thus politics rather than science that has let down the starving and malnourished people of sub-Saharan Africa and Asia.

The facts of food maldistribution are demonstrated by the bald figures: since 1950 world food production has increased by an average of 2.4 per cent a year while population has risen by 1.9 per cent; the increases in supply greater than in demand have consequently led to a steadily falling real price for food on world markets during the same period. Grain, meat and dairy product surpluses pile up in the developed countries, while, in the less developed, more than 500 to 800 million people, in any given year, starve or suffer serious and continuous malnutrition.

What are realistic figures for the population which will have to be fed in the future? Under the high variant of the UN's World Population Prospects 1990 estimates[27] of 1991, there would be an annual growth rate of 1.9 per cent a year in 1990–95 with a decline to 1.4 per cent in 2020–25 resulting in world population of 9.4 billion in 2025 – an annual addition to world population of 128 million a year in the 2020–25 period.

In the UN's low variant, the growth rate would have resumed decline after 1985 and would reach 0.6 per cent in 2020–25 and result in a population of 7.6 billion by 2025.

The medium UN variant estimate foresees an annual growth rate of 1.7 per cent per annum to 1995 and then the resumption of a downward trend to 1.6 per cent in 1995–2000, 1.5 per cent in 2000–2005, 1.2 per cent in 2010–15 and 1 in 2020–25 – population would reach 6.3 billion in 2000 and 8.5 billion in 2025. This medium variant assumes an annual addition

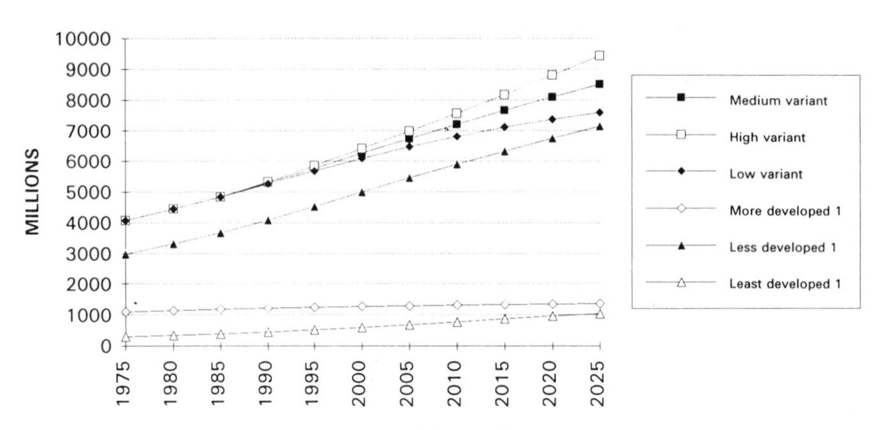

Figure 10.4 UN World population projections

of 83 million in the 2020–25 period. The low variant has already been shown to be a serious underestimate by the observed growth in population in the early 1990s, while it is generally considered that the high variant would be unlikely for the stark fact that if it did begin to develop, rising death rates caused by inadequate resources in Africa would very rapidly cut population back to a lower level.

The expansion of population in Africa will be the biggest problem under any variant: under the medium fertility estimate the population of Africa will stabilise at 3.2 billion – five times the 1990 level – taking till 2085 for Africa to reach 90 per cent of this size. Africa will represent 23 per cent of world population in 2050 and 27 per cent in 2150 (compared with 12 per cent in 1990). The medium variant would appear to anticipate the most likely pattern of events – not just because it is the medium, but because it most closely relates to events so far and reflects a realistic assessment of the impact of the physical and political checks on population expansion – particularly in Africa and the most rapidly developing parts of Asia.

There is, however, a more optimistic scenario: that birth control will be more widely and efficiently adopted, particularly in Africa, and that world population increases will tend to stabilise instead of rising with the prospect of an overall stabilisation of population by the mid-twenty-first century. This was the major conclusion of a recent important study of population change by Stanley Johnson. He suggests that change in reproductive attitudes could mean that fertility rates could decline faster than expected.[28] What is certain, however, is that population will continue to increase for the next fifty years and there will undoubtedly be more mouths to feed.

The problem is and will remain acute in Africa. The region currently needs 14 million tonnes more grain each year than it is able to produce for itself. Population of the African continent is expected to grow by 3 per cent a year over at least the next twenty years; African food production is only increasing by 2 per cent a year. It is clear, therefore, that Africa will have an import need of 50 million tonnes of grain equivalent by 2000. Whether it will be able to afford to import such a quantity – and this is unlikely – depends principally on a substantial increase in aid. Such a quantity could of course be supplied from present levels of world grain output if the necessary finance for such imports were forthcoming.

What would, however, put the world food economy under serious strain would be a very substantial increase in food demand – effective demand backed by money – from the east Asian region of the world. Such a probable increase in demand could also be expected to take the form of increased desire for meat and other livestock products. This latter feature of such increased demand from the dynamic, developing economies of an increasingly wealthy part of the world is what will really put the world food economy under strain. As people become richer – whether Europeans or Asians – they demand a higher-quality diet, and quality in this context means the substitution of meat and dairy products for basic carbohydrate foods. Since seven times as much grain is needed to produce meat as would be needed if the same food requirement were to be eaten simply as grain, increased Asian demand will stimulate a much greater grain demand than would be needed if it were simply a matter of satisfying food demand for an increasing world population.

Current increases in grain production, if they were to continue and population also to increase at the current trend, would result in an average per-capita grain supply of 247 kg by 2000. By 2050, with the same trend, increases on the average would be 244 kg a head. These averages are theoretically adequate to feed the entire world population – but not if the consumption of meat and dairy products also increases. The International Food Policy Research Institute sees an increase in meat-eating in the most rapidly developing countries as the major threat to overall food supply – if there is not an increase in the annual increase in world grain production.[29] Even if there were no increases greater than present trends in world grain supply and no increase in meat-eating in east Asia from current per-capita levels of around 200 kg a year to even the European level of 500–600 (or even worse, the US level of 800 kg), there would be a real world food shortage rather than merely a deepening of the current distributional problem.

To say that food supplies will be adequate in the future is not to say, however, that coaxing the extra production which an increased – and a more prosperous population – will demand will necessarily be easy. The developed world is becoming aware now of the 'hidden costs' of too much of the modern food production industry: nitrate contamination, pesticide residues, rural depopulation and even desertification of once fertile food growing areas. In the recent past these 'external costs' of modern agribusiness were little recognised and certainly were not accepted as barriers to increased production of grain, meat and other animal products.

Agronomists today are, however, only too well aware of the disadvantages of some modern agricultural techniques, but they are also aware that the concern of the people to prevent further damage to the environment by agriculture will impose increasing restraints on the application of the shortest cuts that science might have offered to increased food production. The dilemma was well presented in a recent report from one of the world's major food industry authorities, the US department of Agriculture.[30] The Department is cautious about the sustainability of modern agriculture:

> 'Increased reliance on petroleum energy, fertiliser and pesticides, deep-well irrigation and the farming of marginal land has contributed to environmental problems,' says the Department. 'Future technological advances will have to focus on ways to increase productivity while minimising harmful effects on the environment.'

The implication is, however, that the world's agricultural productive capacity still exceeds the propensity for population and food consumption to increase: 'Even if productivity growth slowly declined from the historical 1.9 per cent to 1 per cent by the year 2040, world average per capita consumption could still be maintained.'

This is demonstrated by the fact that the reduced output of grains and other major food commodities in North America as a result of the 1988 drought – which reduced the US grain harvest by almost a third – was more than made up by the increased production in other areas of the world. If global warming were to make it more difficult to maintain yields in what have come to be regarded as the traditional, major grain-growing areas of the world, increased yields in areas affected by higher temperatures would tend to make up for any shortfall.

The USDA study indicates that the world still has large reserves of under-exploited agricultural productive capacity. This could be brought into play not necessarily by the increased use of fertilisers and crop chemicals, but by the greater use of biotechnological and environmentally more compatible methods. Of course it takes time – more than the average surplus-to-shortage cycle in world food markets – to bring new technology into commercial use, but there are now major innovations which will contribute to increased food production in the future without necessarily demanding more land to be brought into production or the use of more environmentally harmful inputs.

Among these are the evolution of crop varieties with improved drought and disease resistance and with shorter growing seasons that will expand the geographical range of crops (see chapter 9). Scientists are now developing crops which will be able to fix atmospheric nitrogen on a greater scale and in more difficult conditions than the legumes used by European farmers since the sixteenth century to improve the fertility of their soil. It is also expected that new methods of plant protection will reduce toxic residues, while animal growth promotants will increase feeding efficiency and cut the cost of producing meat and dairy products.

Even without the application of these innovations, improved farm management and husbandry can still increase output in many parts of the world. Almost 80 per cent of the increase in world food production in the last three decades has come from increased productivity rather than from the increase in farmed land area. This trend is likely to continue.

But world economic growth will undoubtedly make new demands on agriculture. Rising incomes will mean demand for higher quality diet leading to the need for greater livestock production and therefore greater grain output. There is also the possibility that productivity may not increase as much as expected and that population growth will not decelerate as much as currently projected.

The problem of possible longer-term short supply is likely to be concentrated in the non-industrialising less developed countries. It is therefore likely that producing food for a growing world population will not be the main issue.

The main conclusion to be drawn from the currently available evidence is that world food demand is likely to increase, that production will continue to increase, but may not increase enough to meet the changing pattern of food demand. If environmental restrictions and possible climatic change are not met with adequate technological response, supply will not increase enough or fast enough, to meet all increased demand, and food prices will therefore tend to increase over the longer term. The important conclusion is that, though food may cost more if the scientific response to what will undoubtedly be new conditions for world agriculture is inadequate, in the longer run production will still keep pace with increasing demand.

From a European point of view, the importance of all these analyses is that they indicate that there is little justification for the European countries – inside or outside the European Union – to sustain food production at levels way beyond the level necessary to feed their own people. Indeed, given the European lack of comparative advantage in the production of such essential foods as animal feed, vegetable oils and certain types of breadmaking wheat, there is much to be gained in economic and consumer choice terms from reducing production below EU domestic consumption levels and from importing the resulting deficit from more efficient producers abroad.

Even if the more pessimistic prognosis on world food production capacity from the Meadows' more recent work is accepted, it is clear that west Europeans will be able certainly to continue to produce enough food to feed themselves, or, alternatively, to produce the proportion of their food requirement they desire and to be sure of being able to import any planned deficit. The evidence gathered in this book suggests that, even with the physical and legislative constraints on the use of crop and animal production inputs likely to be imposed during the next two decades, a sustainable agriculture is likely to evolve which will easily be capable of feeding Europe.

Inevitably, however, there will be those who argue that increasing EU food production, at whatever cost, will be justified to meet inevitably increasing world food demand and, in particular, food aid demands.

Much is made of the obvious and continuing need for food by the too often burgeoning populations of the least developed countries of the world as a justification for the maximisation of agricultural production in Europe and the rest of the developed world. If the more dismal analyses examined in this chapter were to prove correct, then there could be expected to be a growing food deficit in some countries of the world. Even with the more optimistic prognoses, it is clear that there will be a continuing food deficit in the poorest countries of Africa.

Why not, then, encourage European farmers to carry on producing more wheat, beef and milk powder than their own fellow citizens can eat and ship the surplus direct to the starving of sub-Saharan Africa? This justification for maintaining high levels of excess production is frequently proposed by farmers representatives.[31] The food aid need for emergency relief and medium term projects is currently put by the World Food Programme organisation[32] at 20–25 million tonnes of grain equivalent a year. The European Union provides just over 1.6 million tonnes of cereals and 100,000 tonnes of milk powder and other dairy products; given its share of world GNP, it ought to be providing substantially more. Should it, however, provide that aid in the form of farm commodities or money?

If the less optimistic projections of food production to population ratio are proved correct, then the need for food commodity aid could rise to 35–40 million tonnes by the year 2020. The ability of the world's farmers to provide such quantities can be judged from the fact that it is equivalent to less than the current combined exportable surplus of the United States and the European Union. Clearly, the EU's contribution would have to increase substantially. Providing this aid in the form of actual agricultural commodities is unlikely to be the most efficient way of meeting this moral commitment for the rich to look after the poor – even if the governments of the Union decide to take on any such commitment.

Given the relative costs of production of additional quantities of grain and other food in Europe, it is probable that a much more efficient way for the EU to meet this commitment would be to provide cash for the purchase of additional food aid commodities. Probably of more long term effectiveness would be the provision of cash for longer-term food production projects in the needy countries.

There are, however, powerful pressures for fulfilling food aid commitments in the form of agricultural commodities – and these pressures come not only from the farm lobby, but also other sections of society who see provision of relief in terms of bags of grain and vacuum packs of powdered milk as being more immediate and tangible and, above all, 'cheaper' than providing money. After all, it is argued, aid in the form of actual food stands more chance of reaching the needy than cash grants that could fall into the hands of corrupt government officials.

As one detailed analysis of the methods and relative efficiency of food aid[33] puts it:

> The fact that food aid has a lot of popular support, being considered as the friend of the donor country's farmers, while financial aid is often very unpopular being considered as the enemy of the donor country's taxpayers, is bound to weigh heavily in deciding to earmark part of the aid programme for food aid.

This argument ignores two important points: first, that the taxpayers of the donor country certainly pay for commodity-form aid – if the country supports its agriculture, as the EU certainly does; second, the costs of production may be so high that the government (on behalf of the aid agency) has to pay more for domestically produced grain than if it purchased it on the international market. For countries and regions with high costs of production, there is no doubt that each dollar of aid cash is most efficiently utilised by buying food aid commodities in the cheapest market.

Only if the world should reach the unlikely state where it has an absolute food shortage, rather than as now a distributional problem, would it be justified to encourage European farmers to maximise production in order to meet a world food shortage. The evidence on world agricultural production presented in this chapter would suggest that such an eventuality is unlikely.

THE 'NEW AGRICULTURE'?

It is clear that the European agriculture industry of the early years of the next century will be very different from the intensifying and specialising industry which became the dominant pattern in the late 1980s. Environmental and social considerations will, by the end of the 1990s, come to dominate agricultural policy formulation. Almost all of the production of major commodities will be subject to restriction from a combination of budgetary cut-backs, international agricultural trade agreements and the limitations imposed by new environmental and social policies.

What were regarded as the major forces driving agricultural policy formulation in the late 1980s and early 1990s – the taxpayer cost of funding agricultural policies, the need to improve the 'market efficiency' of the agricultural sector and the desire to eliminate the wastes of over-production, will be much less significant.

By the late 1990s the international agricultural trade agreements which were hammered out in the GATT during the 1986–93 period will have imposed controls on European agriculture that will have certainly stabilised production of cereals, dairy products and beef. These commodities were the mainstays of the old, production-preoccupied, expansionary farming which dominated the post-war decades. By the late 1990s all will be subject to production quotas – either *de jure* or *de facto* in the form of subsidy limitations which, given the average cost levels of the European farm industry, will have the same effect as actual legal limitations on physical output. Compliance with stringent environmental requirements will form an important integral part of the agricultural support system.

It can be assumed, somewhat paradoxically, that despite the preoccupations of policy-makers in the 1980s with making agriculture 'more subject to market forces', agriculture will have become more rather than less regulated. The overwhelming trend of thinking on agricultural policy in the early 1990s was that 'agriculture policy was too important to be left to agriculturalists'. The predominance of this attitude will lead inevitably to the increasing input of environmental principles into agricultural policy-making with increasing paramountcy not only of society's desire to prevent farming from further polluting the soil, the water and the atmosphere, but also to mould agriculture policy so as to ensure that the countryside and its guardians also increasingly become the providers of 'environmental services'.

The EU-rigged and -manipulated markets for farm commodities will become less and less important in maintaining farmers' incomes. While a large section of the agriculture industry will still gain its income from the sale of the basic food commodities, this produce will be sold at very close to world prices – with little subvention from the EU or national funds. In order to ensure adequate incomes, farmers will still draw direct subsidies from the state (including the EU super-state) to compensate for the very much less intensive methods which they will be forced to practise.

Failure to comply with limitations on the application of nitrogen and phosphates to the land, restrictions on the disposal of livestock effluent and controls on the use of pesticides – all likely to be embodied in 'land management contracts' between farmers and the authorities – will result in non-payment of environmental subsidies which will have largely replaced the market-support subsidies of the past as the main and necessary supplement to farm incomes.

Despite this apparent limitation on the European Union's agricultural production, new scientific developments, mainly in the realms of biotechnology, will sustain still high levels of agricultural production. New plant varieties bred for insect and disease resistance will be teamed with low-concentration but highly effective herbicides to produce still larger crops of high-quality wheat and other commodities. These crops will be genetically engineered to produce more exactly the qualities demanded by the millers and other processors of food. At the same time, a growing number of farmers will be producing very high quality crops of fruit, vegetables, meats, dairy products and first-stage processed foods (such as sausages, patés, cheeses, preserves and other quality 'on-farm' products) which will be demanded by an increasingly critical clientele of direct 'farm-gate shoppers'. There will thus still be the 'two agricultures' which were developing in the 1980s: the bulk food producers and the organic and near-organic quality producers. The difference between the two periods is, however, that even the majority, bulk-producers, will be growing a higher-quality product than they would have done twenty years previously – and in a more environmentally tolerable manner.

Internationally, the European Union will also be operating in a more acceptable manner – acceptable, that is, to other agricultural exporting nations, as well as to its own consumers and taxpayers. By the second decade of the next century, the European Union will be a common market of at least twenty-one countries. Its agriculture will include not only the industries of most of the ex-EFTA countries, but also those of the central European countries. There will have been a switch in the pattern and emphasis of agriculture, with much of the bulk grain and other arable products being grown in the plains of central Europe – eastern Germany, Hungary, Czechoslovakia – as well as in the prime grain-growing areas of northern France and Spain, and the quality products being grown in concentrated areas close to the large conurbations of northern Europe.

There will by then have been a much more marked 'division of labour' among the farmers of Europe as well as a more efficient exploitation of comparative advantage. The removal of technical barriers resulting from the application of the principles of the EU's integrated market since 1993 will have meant the removal of the technical barriers to trade between the agriculture industries of the different EU countries. Closer monetary union – under the guise of the principles set out in the 1992 Maastricht Treaty or some other agreement – will by then have eliminated the currency instabilities which formed such a major obstacle to free European trade in food products and the exploitation of technical efficiency in the 1970s and 1980s.

This is likely to mean that those areas which produce a particular product best will tend to produce most of it – 'best' in this context will mean not only the highest quality, but 'best' in the sense of least use of resources and least strain on the environment. Thus it is probable that cereals and oilseeds will be grown in the north of France and central Europe. Dairy products, beef and lamb will be produced on the grasslands of the north-west (the British Isles, Denmark and Sweden), pigs and poultry close to the 'grain baskets' of the central plains and the bulk of the fruit and vegetables under the Mediterranean sun of the southern countries of the Union. Thus, nearly fifty years after its inception, the European Common Agricultural Policy may achieve the objectives originally envisaged by its founders.

NOTES

1 WHY SUPPORT FARMERS?

1 Knudsen, O. and Nash, J. *Redefining the Role of Government in Agriculture for the 1990s*, World Bank, Washington, DC, 1990.
2 *Agricultural Policies, Marketing and Trade: Monitoring and Outlook*, OECD, Paris, 1993.
3 For a detailed analysis of the way in which the agricultural minority is able to ensure protection and support from the urban majority, see Howarth, R. W., *Farming for Farmers? A Critique of Agricultural Policy*, Institute of Economic Affairs, London, 2nd edition, 1990. Howarth is also good on the corruption of the agricultural economics profession by its employment by agricultural ministries and farm lobby groups.
4 Averyt, W. F., *Agro-Politics in the European Community*, Praeger. New York, 1977. Averyt, an American research scholar, wrote this book after spending an extended period working with and within the COPA organisation in Brussels during its most influential period.
5 *Agricultural Situation in the European Community*, Annual Publication of the European Commission, Brussels. See annual issues 1975–95. Published by the Statistics Office of the European Community, Luxemburg.
6 *Agricultural Policies, Markets and Trade – Monitoring and Outlook* (Annual). Various issues 1988–95, Organisation for Economic Cooperation and Development, Paris.
7 See, for example Koester, U. and Tangermann, S., in *Agricultural Protectionism in the Industrialised World*, ed. F. H. Sanderson, *Resources for the Future*, Washington, DC, 1990.
8 *Reflections on the Future of the Common Agricultural Policy*, EC Commission, Brussels, 1991.
9 *Farming without Subsidies: New Zealand's Recent Experience*, Sandrey R. and Reynolds, R., Ministry of Agriculture and Fisheries, Wellington, NZ, 1990.
10 Ricardo, D., *On Protection in Agriculture*, 1822.
11 *Report and Recommendations on Agricultural Support*, Ministry of Agriculture, Stockholm, 1990.
12 For modern evidence supporting the Ricardian theory in the wake of the EU's 1992 reforms which built new official guarantees of support into farm incomes, see Davidson, G. and Asby, C., *UK Cereals 1993–94: the Impact of CAP Reform on Production Economics and Marketing*, Department of Land Economy, University of Cambridge, 1995.

13 Franklin, M. and Ockenden, J. 'European Agricultural Policy – Ten Steps in the Right Direction', Briefing paper, Royal Institute of International Affairs, Chatham House, Nov. 1994.
14 'The Development and Future of the Common Agricultural Policy' (based on COM(91) 100 and COM(91) 258, European Commission, Brussels, July 1991.

2 DEVELOPMENT OF THE CAP

1 Franklin, M., *Rich Man's Farming: The Crisis in Agriculture*, Royal Institute of International Affairs/Routledge, London, 1988.
2 For a detailed description and analysis of this horrific episode, see Pyke, M., *Man and Food*, World University Library, London, 1970.
3 To see the Franco-German role in the creation of the CAP as merely a means of facilitating the exchange of French agricultural commodities for German industrial goods is a gross over-simplification – mainly because it ignores German desires and objectives in the agriculture and food industry sectors.
4 Wehler, Hans-Ulrich, *The German Empire*, chs 1 and 2 on the development of German attitudes to agriculture policy in the nineteenth century. Berg, Oxford, 1985 (trans.).
5 Hardach, Gerd, *The First World War 1914–1918*, Ch. 5, 'Food Supply in Wartime', *Pelican History of World Economy in the 20th Century*, London, 1987 (original German edition: Deutscher Taschenbuch Verlag GmbH, 1973).
6 Dewey, P. E., *British Agriculture in the First World War*, Routledge, London, 1989.
7 Hendriks, G., *Germany and European Integration. The Common Agricultural Policy: An Area of Conflict*, Berg, Oxford, 1991.
8 Mansholt, S., *Green Europe – The Future for Europe's Farmers*, EEC Commission, Brussels, 1969.
9 See Charlton, M., *The Price of Victory*, British Broadcasting Corporation, London, 1983.
10 See Neville-Rolfe, E., *The Politics of Agriculture in the European Community*. Policy Studies Institute, London, 1984. Ch. 4 for one of the few perceptive analyses, in English, of French attitudes to agriculture policy.
11 *Livre blanc de l'agriculture française*, APPCA, Paris, 1965.
12 The term 'agribusiness' is used throughout this book to distinguish between agricultural activity based on normal, generally accepted commercial principles of profit and loss and peasant-based survival agriculture, which is little concerned with normal commercial principles.
13 Eurostat, Luxemburg, Annual Reports on the Structure of Farm Holdings in the European Community, various issues, 1975 to 1993.
14 Vedel Report – Perspective à longue terme de l'agriculture française (1968–1985), report on committee proceedings chaired by M. le doyen Vedel, La Codumentation Française, Paris, 1970.
15 The complexities of Germany's coalition governmental system and the relationships between the parties are well described in Derbyshire, I., *Politics in West Germany, from Schmidt to Kohl*, Chambers, London, 1987.
16 See Perry, P. J., *British Farming in the Great Depression 1870–1914: An Historical Geography*, David & Charles, Newton Abbot, 1971.
17 For agriculture the depression began sooner than for the industrial sector – in Britain principally because of the removal of the temporary support and protection introduced at the end of the First World War.
18 Cooper, A. F., *British Agricultural Policy 1912–36: A Study in Conservative Politics*, Manchester University Press, Manchester, 1989.

19 Pollard, S., *The Development of the British Economy 1914–1980*. 3rd edition, Edward Arnold, London, 1983.
20 Donaldson, F., Donaldson, J. G. S. and Barber, D., *Farming in Britain Today*, Pelican, London, 1972.
21 Whetham, E. H., *British Farming 1939–49*, Nelson, London, 1952 (actually covers the period from the end of the First World War).
22 Hammond, R. J., *Food: The Growth of Policy*, UK Civil Series, HMSO, London, 1951.

3 HOW THE EUROPEAN UNION SUPPORTS AGRICULTURE

1 Court of Auditors of the European Community, *Annual Report on the EC Budget*, Luxemburg, 1991.
2 International Wheat Council (now International Grains Council), *Monthly Reports on the International Wheat and Cereals Market*, 1978–92.
3 See *Agra Europe weekly newsletter* on EU agriculture policy and trade (English edition, London), 1975–92.
4 Ibid.
5 These total figures are given in US dollars because the EC's unit of account system changed radically in 1978–79 when the European Currency Unit came into operation – comparison between the pre- and post-1979 periods is therefore not possible in units of account.
6 European Commission, *Annual Report on the Situation in Agriculture*, 1972–93.
7 Court of Auditors, *Annual Reports on the EU Budget*, 1989–94. Luxemburg.
8 See the weekly EC agriculture policy bulletin *Agra Europe (Agra Europe (London) English edition)*: many reports throughout the 1980s.
9 European Commission, *The Rural World: Report on the Income and Social Situation of Rural Areas*, Brussels, 1989.
10 Even after the 1992 reforms this dominance of agricultural spending in the total EU budget is likely to remain; it is expected to continue to be more than 50 per cent until at least the end of the 1990s.
11 Franklin, M. *Rich Man's Farming*, Royal Institute of International Affairs/ Routledge, London, 1989.
12 EC Court of Auditors, *Report on the Payment of Dairy Export Refunds*, April 1992.
13 *Agra Europe* (German edition), Bonn, April 1973.
14 See *Agra Europe (London) Weekly Newsletter* – intervention stock reports, 1978–93.
15 See, in particular, *The Management and Control of Export Refunds*, Court of Auditors Special Report, Luxemburg, 1990.
16 Preamble to *Proposals for the Reform of the CAP*, EC Commission, Brussels, July 1991.
17 *Agricultural Situation in the Community*, Annual reports 1986–95, European Commission, SOE, Luxemburg.
18 The weekly European agricultural policy newsletter *Agra Europe* (English edition), various editions 1988–92.
19 *Agra Europe Newsletter* (London), April 1989.
20 *Agricultural Policies, Markets and Trade. Monitoring and Outlook* (annual), 1987–94, Organisation for Economic Cooperation and Development, Paris.
21 Apart from the OECD annual reports on agriculture policies, already referred to, the most important are: (1) *Agricultural Policies in the European Community:*

Their Origins, Nature and Effects on Production and Trade, Policy Monograph No. 2, Bureau of Agricultural Economics. Canberra, Australia, 1985; (2) *The Costs of the Common Agricultural Policy*, Buckwell, A. E., Harvey, D. R., Thomson, K. J. and Parton, K. A., Croom Helm, London, 1982; and (3) *Global Effects of Liberalising Trade in Farm Products*, Anderson, K. and Tyers, R., Trade Policy Research Centre, London, 1991.

22 *The Costs of the Common Agricultural Policy*, Buckwell, A. E., Harvey, D. R., Thomson, K. J. and Parton, K. A., Croom Helm, London, 1982.

23 *Global Effects of Liberalising Trade in Farm Products*, Anderson, K., and Tyers, R., Trade Policy Research Centre, London, 1991.

24 The US dollar is again used in these calculations to overcome the confusion created by the EU's 'agri-monetary' system.

25 *Consumers and the Common Agricultural Policy*, National Consumer Council, HMSO, London, 1988.

26 *Impact of Danish EC Membership on Dietary Patterns*, Forbrugerradet, Copenhagen, 1987.

27 GATT and the Reform of Agriculture Policy in OECD Countries, Hewitt, J., OECD, Paris. Paper to Agra Europe Outlook Conference, London, February 1994.

28 *Agricultural Policies Marketing and Trade: Monitoring and Outlook 1995*, OECD, Paris.

4 THE AGRICULTURAL REVOLUTION

1 European Commission, Brussels: *Agricultural Situation and Outlook, Annual Reports*, 1970–80.

2 Milk Marketing Board (UK), *EEC Dairy Farming Yearbooks* 1973–82.

3 *Report on the Structure of the EC Dairy Industry*, European Commission, Brussels, 1989.

4 Ibid.

5 *Agricultural Situation in the Community*, Annual, years 1970–90, European Commission, Brussels.

6 US dollars are used rather than ECU for ease of international comparison.

7 *Differences in Agricultural Research and Productivity among 26 Countries*, Frisbold, G. B. and Lomax, E., US Department of Agriculture, Agricultural Economic Report No. 644, 1991.

8 *Report on the Structure of the EC Dairy Industry.*

9 International Technology Transfer in Agriculture, Anderson, M., *USDA Agriculture Information Bulletin* No. 571, August 1989.

10 *Agriculture – Statistical Yearbook*, Eurostat, issues 1970–95, Statistics Office of the European Communities, Luxemburg.

11 Farm Structure Surveys 1966/67–87, European Commission. Eurostat SOE, Luxemburg; plus unpublished Commission data, 1988-94.

5 GOING TO THE TRADE WARS

1 Though the 1992 CAP reforms and the 1994 GATT agreement are intended to reduce surpluses and subsidised exports, both are likely to persist into the next century. A substantial differential between EU prices and world prices will persist.

2 For a detailed analysis of this development, see Buckwell, A., The CAP and World Trade, in *The Common Agricultural Policy and the World Economy: Essays*

in Honour of John Ashton, ed. Ritson and D. Harvey, CAB International, Oxford, 1991.

3 Buckwell, The CAP and World Trade, p. 228.

4 General Agreement on Tariffs and Trade, *World Trade in Food in the 1980s*, Geneva, 1991.

5 The relationship between the export of the Community's surplus 'C' sugar and the world market price is explained in *Effect of a GATT Uruguay Round Agreement on the European Community and World Sugar Markets*, EPA Associates, Brussels, 1992.

6 Agricultural Policies in the European Community: Their Origins, Nature and Effects on Production and Trade, Policy Monograph No. 2, Bureau of Agricultural Economics, Canberra, 1985.

7 *Agricultural Situation in the Community, 1995*, European Commission, Brussels, SOE, Luxemburg.

8 Throughout this chapter, data on production, export quantities and, particularly, world prices and EC and US export subsidies are taken from various issues of the weekly European agricultural policy and trade newsletter *Agra Europe* (UK edition), in the 1970–92 period.

9 The impact of the CAP on British farming between EC accession in 1973 and 1980 is well documented and analysed by the UK Ministry of Agriculture's Chief Economist in Capstick, C. W., The British Agricultural Policy under the CAP, in *The Common Agricultural Policy and the World Economy: Essays in Honour of John Ashton*, CAB International, Oxford, 1991.

10 Capstick, C. W., The British Agricultural Policy, p. 86.

11 Morgan, D., *Merchants of Grain*, Viking Press, New York, 1979.

12 With the mental and political flexibility which ensures its survival and dominance of the world grain trade, Cargill abandoned its opposition to the CAP once the 1972–3 enlargement became inevitable. As a major beneficiary of Brussels agricultural trade handouts, the company became one of the most avid supporters of the CAP in the 1980s.

13 Halmar, M., Meyers, W. H. and Thamadoran, R., Impact of EC Policies on US Export Performance: Confrontation or Negotiation? *United States Policy and European Agriculture*, The Curry Foundation, Associated Faculty Press, Millwood, NY, 1985.

14 Australian Bureau of Agricultural and Resource Economics, Canberra, US Grain Policies and the World Market, Policy Monograph No. 4, 1989.

15 The North African Grain Market, USDA Staff Paper, Washington, DC, 1989.

16 Matthews, A., *The Common Agricultural Policy and the Less Developed Countries*, Gill & Macmillan, Dublin, 1985. See also Halmar *et al.*, Impact of EC Policies.

17 Lingard, J. and Hubbard, L., The CAP and its Effects on Developing Countries, *The Common Agricultural Policy and the World Economy*, CAB International, Oxford, 1991.

18 Halmar *et al.*, Impact of EC Policies.

19 US Department of Agriculture, *Impact of the GATT Uruguay Round on Agricultural Trade*, Washington, 1994.

20 Lowe, G. and Roberts, I., The Central Role of US Grain Stocks, *Quarterly Review of the Rural Economy*, 10 (4) December 1988), ABARE, Canberra.

21 Legras, G., Reasons for Not Following US Line on CAP Reform, Paper to CEPS working party on the future of the CAP, Brussels, 1990. Reported in *New Directions for European Agricultural Policy*, Arnold, R. and Villain, C., Centre For European Policy Studies, Brussels, 1990.

22 The so-called Draft Final Act for the Uruguay Round of Multilateral Trade Negotiations in the General Agreement on Tariffs and Trade, Geneva, 20 December 1991.

23 The members of the Cairns Group are: Argentina, Australia, Brazil, Canada, Chile, Colombia, Fiji, Hungary, Indonesia, New Zealand, Malaysia, the Philippines, Thailand and Uruguay.
24 See Part Two I, The Development and Future of the Common Agricultural Policy, Document COM(91) 100 and COM(91) 258, EC Commission, Brussels, July 1991.
25 Anderson, K. and Tyers, R., *Global Effects of Liberalising Trade in Farm Products*, Trade Policy Research Centre, London, 1992.
26 US Department of Agriculture, Analysis of the Effect of the GATT Agreement on US Agricultural Trade, USDA, Washington, DC, 1994.
27 See Chapter 6.
28 See note 25.
29 See note 25.
30 See, for example, *International Comparisons of Costs of Wheat Production in the EC and the United States*, Bureau, J-C., Butault, J-P and Hogue, A., USDA Staff Report No. 9222, Washington, DC, 1992.
31 Aart de Zeeuw, Blueprint for a GATT compromise, *European Affairs* (April/May 1991).

6 THE GREAT REFORM?

1 For examples of this position see: European Agriculture Policy – Ten Steps in the Right Direction, Franklin, M. and Ockenden, J., Briefing paper, RIIA, Chatham House, November 1994, and Expert Group Report to the UK Ministry of Agriculture Fisheries and Food on the Future Development of the CAP, July 1995.
2 Commission reform proposals: The Development and Future of the Common Agricultural Policy, DOC, COM(91) 100 and DOC, COM(91) 285, Brussels, July 1991.
3 Council agreement on CAP reform May 1992: the most important legislation in this reform package affects the cereals and arable sectors and is contained in, principally, Regulation 1765/92 (which deals with compensatory payments and set-aside). The whole package for every commodity sector is summarised in The New Regulation of the Agricultural Markets, *Vademecum: Green Europe Newsletter* No. 1, 1993, published by European Commission (DGVI), Brussels.
4 In 1994–95 this was reduced to 12 per cent and in 1995–96 to 10 per cent of basic arable area.
5 The average of the five years 1988–92, excluding the highest and the lowest.
6 Policy Impact Analysis (a detailed study of the impact on farmers' incomes of CAP reform and GATT in most member states of the EU and covering all the main commodities), EPA Associates/Agra Europe (London) Ltd, Tunbridge Wells, 3rd edition, 1995.
7 Herlihy, M. T., Effectiveness of Acreage Control Programs in the United States: Implications for the European Community, Paper to Conference, Calabria, June 1993.
8 Home Grown Cereals Authority, London: EU Cereal Balance Sheet, *Weekly Digest*, 9 January 1995.
9 Home Grown Cereals Authority, London: The GATT Agriculture Agreement, Marketing Note 1.8.94. See also Guyomard, H., INRA, The Implications of Political Decisions on the EU Agriculture Industry, Paper to ECPA Meeting, Brussels, June 1994.
10 Meat and Livestock Commission, UK: CAP Reform – the Challenge of Change, November 1992.

11 US Department of Agriculture, Washington, DC: The Relationship of the 1992 CAP Reforms to the 1994 Uruguay Round Agreement, Internal staff report, unpublished.
12 See EU Commission budget estimates 1995, and Home Grown Cereals Authority, London: EU Farm and Cereal Budgets 1995, Marketing Note, 20 March 1995.
13 Abstracted from *Policy Impact Analysis* – an analysis of CAP reform and GATT on the EC's common agricultural policy, EPA Associates/Agra Europe joint publication, March 1993.
14 French Government Memorandum on the compatibility of a GATT agreement with the 1992 Reforms of the CAP, Paris/Brusels, November 1993. See also International Commission Directorate General VI (Agriculture) background note on the *incompatibility* of CAP with the GATT. Same date, unpublished.
15 Compatibility of the 1993 Breydel Agreement with CAP Reform, European Commission, published Brussels, December 1993.
16 International Trade Reform: What the GATT Negotiations Mean for US Agriculture, US Department of Agriculture Staff Paper, USDA, Washington, DC, 1990.

7 IMPACT OF THE CAP REFORM AND THE GATT

1 USDA Staff Report, No. 9222, 1992.
2 EC evidence of the ability of EC farmers to produce at the world price?
3 Commission statement on the compatibility of the Blair House accord with the 1992 CAP reform programme, Brussels, December 1992.
4 USDA staff reports on the impact of the 1994 GATT agreement on the CAP, 1994–95.
5 See note 3.
6 See, for example, *UK Home Grown Cereals Authority, Weekly Digest*, February 1993.
7 Herlihy, M. T., The Effectiveness of Acreage Control Programmes in the US: Implications for the European Community, Paper to conference, Calabria, June 1993.
8 Report on Farming in Eastern England 1992–3, Department of Land Economy, University of Cambridge.
9 The Effectiveness of the EC's New Set-aside Programme: An Assessment Based on US Experience and Interviews with EC Farmers, Herlihy, M. T. and Madell, A. L. *Agricultural Trade Conflicts and the GATT*, Boulder, CO: Westview Press, 1994.
10 The system was further weakened in 1995 by the introduction of a facility which allows farmers to 'export' their set-aside liability to other farmers: this allows further movement of set-aside away from the most productive to the less productive land in the Union.
11 European Commission Directorate General VI (Agriculture), monthly and annual crop reports, 1992–95.
12 Preliminary analysis of the economic implications of the Dunkel text for American agriculture, Office of Economics, USDA, Washington, DC, March 1992.
13 Ibid.
14 The Food, Agriculture, Conservation and Trade Act of 1990, Washington, DC, 1990.
15 Changes in Cereals and Dairy Policies in OECD Countries: A Model-based Analysis, OECD, Paris, 1991.

16 This effect is indicated in all computer models and economic assessments which have addressed the issue of the effect on world market prices of the subsidisation of grain exports by the major exporters. Apart from the OECD model drawn upon in this chapter, see, for example: Anderson, K. and Tyers, R., Global Effects of Liberalising Trade in Farm Products, Trade Policy Research Centre, London, 1991; or US Grain Policies and the World Market, Policy Monograph No. 4, Australian Bureau of Agricultural Resource Economics, Canberra, 1989.
17 For summary of these arguments see, for example, Gardner, B. The 1993 GATT Uruguay Round Agreement: Impact on European Community Agricultural Exports. Catholic Institute for International Relations, Seminar paper, London, 1994
18 Economic Research Service, US Department of Agriculture, Washington, DC.
19 Anderson, K. and Tyers, R. Global Effects of Liberalising Trade in Farm Products.
20 See, for example, Policy Impact Analysis: Assessment of the Effect of CAP reform and the GATT on European Agriculture, EPA Associates/Agra Europe, Agra Europe (London) Ltd, Tunbridge Wells, UK, 1995.
21 Agricultural Strategy, 1994. Agricultural Development and Advisory Service, London. See also quarterly international market reports of the UK Meat and Livestock Commission.
22 Ibid.
23 Ibid.
24 European Commission EAGGF budget estimates, 1994 and 1995.
25 SPEL Sectoral Production and Income Model for Agriculture Data for EU Agriculture, 1985–94, Eurostat, Luxemburg.
26 Sectoral Production and Income Model for Agriculture.
27 UK Cereals 1993–94, The Impact of the CAP Reform on Production Economics and Marketing, Geoff Davidson and Carol Asby, Department of Land Economy, Cambridge University.
28 Agricultural Policies, Markets and Trade in OECD Countries, 1994, OECD, Paris.
29 See note 26.
30 Anderson and Tyers, Global Effects of Liberalising Trade in Farm Products.

8 EUROPEAN FARMING AND THE ENVIRONMENT

1 Environment and agricultural policies, Martin Holdgate – paper to conference on agriculture, food and rural policies in the European Union, Chatham House, London, November 1994.
2 Dutch Government Environment Plan to 2000, Ministry of Housing and the Environment, The Hague, 1988.
3 See, for example The Fate of Nitrogen Applied to Agricultural Crops with Particular Reference to Denitrification, Dowdell, R. J., *Philosophical Transactions of the Royal Society*, B296, 1982.
4 Unwelcome Harvest: Agriculture and Pollution, Conway, G. R. and Pretty, J. N., Earthscan Publications, London, 1991.
5 *Agriculture and Pollution*, Royal Commission on Environmental Pollution, 7th Report, Cmnd 7644, HMSO, London, 1979.
6 Intensive Farming and the Impact on the Environment and the Rural Economy of Restrictions on the Use of Chemical and Animal Fertilisers. Study for the EC Commission by TEAGASC (The Agriculture and Food Development Authority), Dublin, 1989.

7 Von Meyer, H. The Common Agricultural Policy and the environment: the Effects of Price Policy and the Options for Reform. World Wide Fund for Nature, CAP discussion paper No. 1, 1988.
8 Ibid.
9 Ibid.
10 European Commission (DGXI) Environment and Agriculture COM (1988) 338 Brussels 1988.
11 Intensive Farming.
12 In EU legislation a 'Directive' is less prescriptive than a 'Regulation'. While providing a body of law to solve a particular problem, it allows considerable leeway to the member state government to interpret in its own way and to mould it into its own legislation.
13 Leuck, D. J., Haley, S. and Liapis, P., The Relationship between Selected Agricultural and Environmental Policies in the European Community. Staff paper, USDA Economic Research Service, Washington, DC.
14 *Official Journal of the European Communities,* L375, December 1991.
15 Koopmans, T. An Application of an Agro-Economic model to Environmental Issues in the EC: a Case Study *European Review of Agricultural Economics,* 14 (2), 1987.
16 European Fertiliser Manufacturers Association, Brussels, *Annual Report,* 1994.
17 Leuck, D. J., Policies to Reduce Nitrate Pollution in the European Community and Possible Effects on Livestock Production, USDA Economic Research Service Staff Report No. AGES 9318, 1993. Calculations on the basis of Koopmans' original research.
18 Liapis, P. Environmental and Economic Implications of Alternative EC Policies, USDA Economic Research Service, unpublished monograph.
19 Leuck *et al.*, Relationship between Selected Agricultural and Environmental Politics.
20 Dutch Government Environment Plan to 2000.
21 Ibid.
22 European Commission. Environmental Policy in the European Community 1990.
23 Environment Commissioner Carlo Ripa di Meana in a speech to a Brussels conference, November 1991.
24 Focus on the CAP, discussion paper, Country Landowners Association, London, 1994.
25 Von Weizsacker, W., Institute for European Environmental Policy, Bonn, Agra Europe Special Report No. 43, London, 1988.
26 Reported in: Environmental and Economic Implications of Alternative EC Policies, Peter S. Liapis, *Journal of Agricultural and Applied Economics,* 26 (1), July 1994. The study uses the SWOPSIM static partial equilibrium net trade model developed by the USDA, by Peter Liapis and other Economic Research Service economists.
27 Expert report to the European Parliament Agriculture Commitee on the Impact of Modern Farming on the Rural Environment, 1992.
28 Ibid.
29 See note 24.
30 Pearce, D., Economic Values and the Natural World, Centre for Social and Economic Research on the Global Environment, Earthscan, London, 1993.
31 Caring for the Earth, a Strategy for Sustainable Living, IUCN/UNEP/WWF, 1991.
32 Farming for a Greener Britain: Farming and the Environment in the 21st Century, Birdlife International and WWF UK: Dixon, J. and Murray, G., Sandy, Beds., UK, 1994.

33 For an objective examination of both arguments, see: Anderson, K. and Blackhurst, R. (eds) *The Greening of World Trade Issues*, Harvester Wheatsheaf, Hemel Hempstead, 1992.
34 Pearce, D., *Blueprint for a Green Economy*, Earthscan Publications, London, 1991.
35 'Sustainable economic growth means that real GNP per capita is increasing over time and the increase is not threatened by "feedback" from either bio-physical impacts (pollution, resource problems) or from social impacts (social disruption).'
36 *The State of Agriculture in the United Kingdom*, A report to the Royal Agricultural Society of England, Barber, *et al.*, 1991.
37 *Alternative Agriculture*, National Research Council, National Academy Press, Washington, 1989.
38 Organic Farming as a Business in Great Britain, Murphy, M. C., Agricultural Economics Unit, University of Cambridge, 1992.

9 THE 'BIOTECH' REVOLUTION?

1 See Fincham, J. R. S. and Ravetz, J. R., *Genetically Engineered Organisms – Benefits and Risks*, Open University Press, Milton Keynes, 1991; or Dixon, P., *The Genetic Revolution*, Kingsway Publications, Eastbourne, 1993.
2 *Use of Somatotropin in Livestock Production*, Sejrsen, K. Vestergaard, M. and Miemann-Soreason, A. (eds), European Commission and Elsevier Applied Science Publications, 1989.
3 Bauman, D. E., Bovine *Somatotropin: Review of an Emerging Animal Technology*, Office of Technology Assessment, Congress of United States, Washington, DC, 1990.
4 EC Commission, Report on the State of the Community Dairy Market, Brussels, 1989.
5 Bu'Lock, J. and Kristiansen, B., *Basic Biotechnology*, Academic Press, London, 1987.
6 Lindsey, K. and Jones, M. G. K., *Plant Biotechnology in Agriculture*, John Wiley & Sons, Chichester.
7 *Field Releases of Transgenic Plants 1986–1992: an Analysis*, OECD, Paris, 1993.
8 *A New Technological Era for American Agriculture*, Office of Technology Assessment, Congressional Board of the 102d Congress, Washington, DC, Government Printing Office, 1992.
9 Ibid.
10 Ibid.
11 Campbell, J. R., The Ongoing Green Revolution: Meeting the Global Food Challenge, in *Global Food Progress 1991*, Hudson Institute, Indianapolis, 1991.
12 Brown, L. R., Facing Food Insecurity, in *State of the World 1994*, Worldwatch Institute, Washington, DC, 1994.
13 Biotechnology, Agriculture and Food, Organisation for Economic Cooperation and Development, Paris, 1992.
14 *A New Technological Era for American Agriculture.*
15 The Impact of Biotechnology on Agriculture in the European Community to the Year 2005, Bureau Européen de Recherches, Study for EU Commission DGVI, Brussels, 1989.
16 *Field Releases of Transgenic Plants 1986–1992.*

10 EUROPEAN AGRICULTURE IN THE TWENTY-FIRST CENTURY

1 *Agricultural Policies, Markets and Trade in the CEECs, the NIS and China: Monitoring and Outlook*, OECD, Paris, 1993.
2 *Agricultural Policies, Markets and Trade in the Central and Eastern European Countries, Selected Newly Industrialised States, Mongolia and China: Monitoring and Outlook, 1995, OECD*, Paris, 1995.
3 The Budgetary Implications of EC Enlargement, Brenton, M. and Gros, B., working document, Centre for European Policy Studies, Brussels, May 1993.
4 Munk, K. J., *The Development of Agricultural Policies and Trade Relations in Response to the Transformation in Central and Eastern Europe*, European Economy – Reports and Studies No. 5, 1994, European Commission, Brussels.
5 Tangermann, S., The CAP in a Growing EC and a Rapidly Changing EC Political Scenario: Implications for US–EC Agricultural Trade Relations, Conference paper, Calabria, June 1993.
6 Brenton and Gros, Budgetary Implications of EC Enlargement.
7 See, for example, report by Henri Nallet and Adrian van Stolk to the EU Commission on the problems of agriculture in CEEC and their relationship to EU agricultural trade, June 1994.
8 *EC Agriculture Policy for the Twenty-First Century*, Report of an expert group of economists for the EC Commission, autumn 1994.
9 This question should always have been at the root of European agricultural policy formulation; it is only the events of the 1980s and early 1990s that has forced it to the forefront.
10 *The Limits to Growth*, Meadows, D. H. *et al.*, Universe Books, New York, 1972.
11 *Beyond the Limits*, Meadows, D. H., Earthscan Publications, London, 1992.
12 FAO (UN Food and Agriculture Organisation), *World Agriculture: Towards 2000*, Rome, 1988.
13 Carr-Saunders, A. M., *World Population*, Oxford University Press, 1936.
14 *Global Food Progress, 1991*, Avery, D. T., Hudson Institute, Indianapolis, 1991.
15 Avery, D. T., World Food Production: Saving the Earth's Wildlife with Farm Chemicals, Conference paper, Brussels, June 1994.
16 Ibid.
17 *Computation of the Absolute Maximum Food Production of the World*, Buringh, P., Van Heemst, H. D. J. and Staring, G. J., Wageningen, 1975.
18 *World Development Report 1992: Development and the Environment*, World Bank, Oxford University Press.
19 GRC Economics, *Report on US Food Production Capacity*, Washington, DC, 1990.
20 *Estimate on Disappearance of Crop Chemicals*, Report by European Crop Protection Association, Brussels, 1990.
21 *Climate Change 1992*, World Meteorological Organisation/United Nations Environment Programme, Intergovernmental Panel on Climate Change.
22 *Climate Change: Economic Implications for World Agriculture*, Kane, S., Reilly, J. and Tobey, J., US Department of Agriculture, Washington, DC, 1991.
23 *The Impact of Climatic Variations on Agriculture*, Parry, M. L., Carter, T. R. and Konijn, N. T., The International Institute for Applied Systems Analysis/United Nations Environment Programme.
24 *Reports on World Resources 1990–91 and 1992–93*, World Resources Institute, Oxford University Press, New York.

25 Various publications, but see particularly: 'Feeding the World in the Nineties', *State of the World 1990*, Brown, L. R. and Young, J. E., W. W. Norton and Co., New York, 1990.
26 How Many People Can Be Fed on Earth? Heilig, G. H., *The Future Population of the World*, ed. Lutz, W., International Institute for Applied Systems Analysis, Earthscan Publications, London, 1994.
27 World Population Prospects 1990, United Nations, New York, 1991.
28 *World Population – Turning the Tide*, Stanley P. Johnson, Graham and Trotman, London, 1994.
29 *Agricultural Productivity in the Third World: Patterns and Strategic Issues*, Report of Seminar, 1993, International Food Policy Research Institute, Washington, DC, 1993.
30 *Alternative Agriculture* (Report of Committee on the Role of Alternative Farming Methods in Modern Production Agriculture), National Research Council, National Academy Press, Washington, DC, 1989.
31 Richardson, D., *Financial Times* regular column on agriculture, 'From a farmer's point of view', 1990–95.
32 World Food Programme, *Review of Food Aid Policies and Programmes*, Rome, 1983.
33 *Food Aid: the Challenge and the Opportunity*, Singer, H., Wood, J. and Jennings, T., Clarendon Press, Oxford, 1987.

INDEX